AF334081

Modern Sawmill Techniques

Proceedings of the first Sawmill Clinic

Portland, Oregon, February 1973

Edited by Vernon S. White

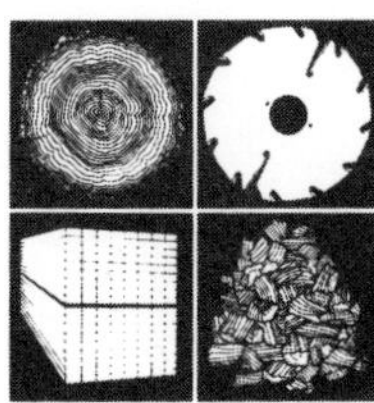

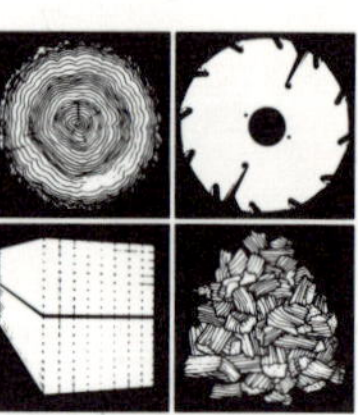

Volume 1 of the Sawmill Clinic Library
A FOREST INDUSTRIES book

Library of Congress Catalog Card Number 73-88045
International Standard Book Number 0-87930-022-1

Table of Contents

Table of Contents

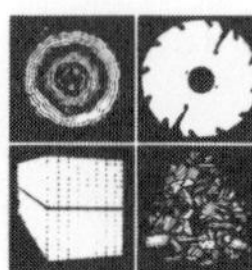

Foreword by the editor

The lumber industry in North America is moving through its industrial revolution.

Until the seventies, the principal spur to technical improvement within our industry was labor cost—or the hope of reducing it.

While raw material cost had been rising slowly for some time, increasing log prices and log scarcity did not threaten sawmill lifelines until 1972. As the tide of demand for lumber swelled, the supply of lumber ebbed. Preservationist pressure locked more North American forests into wilderness. Export demands soared and Japanese ships carried more logs away from North American shores.

Clearly, survival today dictates that our sawmill managers recover every possible board foot of usable lumber from every log they handle. Now, a myriad of conflicting products and systems compete to fill the technological void which the sawmiller must bridge in order to optimize production and recovery.

Miller Freeman Publications, Inc., undertook the job of sponsoring a Sawmill Clinic on February 16, 17 and 18, 1973. The objective was to bring the best and newest operational and management ideas together at one marketplace, thus providing a forum where sawmillers and their industry suppliers could compare, digest, and discuss their common problems and experience.

The formal Clinic sponsors were FOREST INDUSTRIES and WESTERN TIMBER INDUSTRY, two of Miller Freeman's group of natural resource publications, these two specifically serving the logging and sawmilling, plywood and board manufacturing industries of North America.

Twenty-seven presentations were made in a crowded two- and one-half-day program at the Memorial Coliseum in Portland, Oregon. Some 48 booths exhibited the manufacturers' and dealers' products and services. Computers with the software programs to make them work, movies and models of sorters, slabbers, scanners, interfacing devices and a vast array of system concepts were presented by Clinic exhibitors and speakers.

Time was provided between and after each day's program sessions for Clinic registrants to explore the ideas presented by the speakers.

When Miller Freeman's James C. Wallace, vice president and publisher of the company's forest resource division publications, first conceived the idea of a Sawmill Clinic, he asked me as editor of WESTERN TIMBER INDUSTRY, to assemble an Advisory Committee. The purpose of this committee was to discuss the feasibility of the Clinic plan and counsel with us on a possible Clinic program. Many thanks are due its 14 members, listed in this book on page 9.

At that stage of the Clinic's development, we were actually thinking in terms of calling our gathering a symposium or seminar. But C & D Lumber Co.'s forthright boss, E. P. "Bud" Johnson, vetoed that terminology. No, he would never submit to being bored at a symposium or seminar or any such formal, fancy gathering. But he would attend a Clinic—might really learn something at a Clinic. So the first 1973 Sawmill Clinic was there born and named.

The first Clinic, held in Portland, was extremely successful. Originally it was estimated that about 300 to 350 sawmillers would attend. When registration was closed, over 850 industry members had registered. The audience was enthusiastic, serious and attentive. The speakers exceeded even my confident expectations in the wealth of ideas and information they presented.

Obviously, to reproduce the material presented at the Clinic sessions so that it would be available for permanent reference was important.

Consequently, this book was conceived. We hope the Clinic deliberations, reassembled here in book form, will help the industry bridge today's technology gap.

It is our hope also that we can continue to sponsor such clinics in the future when and if sufficient practical new developments, new techniques and new ideas have emerged to warrant their presentation in such a manner.

In fact, at this writing two more clinics are in the active planning stage: one for the sawmill industry in the South and a second one in the West. The latter will include also an extra day for similar clinic sessions devoted exclusively to practical new ideas and techniques in plywood manufacturing.

From these clinics will spring additional proceedings books, such as this one, to capture in permanent form the ideas and techniques presented at these sessions. Thus this book is volume one of what we hope and expect will be a growing library of modern manufacturing techniques for the sawmill, plywood and other forest products manufacturing industries.

Our special appreciation is extended to our speaker/authors who took the time to prepare and present their Clinic papers and to share their knowledge and experience with the Clinic registrants. To a man they have done a superb job of presentation both at the Clinic and later in the tremendous task of adapting their material to this book.

I also want to thank the Portland Chamber of Commerce, the Western Forestry Center and the many people too numerous to mention who assisted with tours and with the Clinic facility requirements, and we appreciate the hospitality extended to our guests.

We had a fine, impressive group of industry exhibits and we appreciate the contribution of the exhibitors who supported the project from the beginning.

Finally, our hope is that you will use this text and find it a valuable reference—of real and practical help as you face the complications of our industry's future health and survival.

Vernon S. White
Editor
Portland, Oregon
August 1973

Exhibitors
1973 Sawmill Clinic
Portland, Oregon

While they are not reflected within the pages of this book, many valuable ideas were gathered by those attending the first Sawmill Clinic during visits to exhibitors' booths and discussions with their representatives. The sponsors of the Clinic appreciate the exhibitors' support of the proceedings. The following were the exhibiting companies:

A-1 Steel & Iron Foundry
Albany International
   Industries, Inc.
American Sheet Metal, Inc.
Archer Blower & Pipe Co., Inc.
Armstrong Mfg. Co.
AutEx, Inc.
Bigelow Machinery Inc.
A. B. Boyd Co.
CAE Machinery Ltd.
Chip-N-Saw, Inc.
Chipper Machines &
   Engineering Corp.
Computer Sciences
   Corp./McGrew Bros.
Cutler-Hammer, Inc.
Dyna-Tech Corp.
Elworthy & Co. Ltd.
Fluid-Air Components, Inc.
Gear Reducer Sales Co.
Graffinberger Valve Co.
Howard-Cooper Corp.
Industrial Machinery
   Erectors Co.
Interlake Steel Corp.
International Business
   Machines Corp.

Irvington-Moore
Jacksonville Blow Pipe Co.
Lloyd Controls, Inc.
Mainland Foundry &
   Engineering Ltd.
Mark 50 Machinery Inc.
H. C. Mason & Associates, Inc.
Microdyne Modular Electronics
   Systems Ltd.
Mill Conversion
   Contractors, Inc.
Moore-Oregon, Inc.
Morrison Knudsen Co., Inc. &
   RDA Inc.
Opcon Inc.
Portland Iron Works
Ramic Corporation
Rens Mfg. Co., Inc.
Salem Equipment Co.
Simpson Timber Co.
T. E. Slanker Company
Soderhamn Machine
   Manufacturing Co.
Stetson-Ross Machine Co., Inc.
Tyrone Hydraulics
U.S. Department of Commerce
Wagner Electronic Products

Advisory Committee
1973 Sawmill Clinic
Portland, Oregon

The success of the first Sawmill Clinic was due in great part to the contributions of its Advisory Committee. It was they who met with the sponsors in July 1972 to assess the feasibility and practicality of such a meeting and who later gave much time and energy working for its success. Committee members suggested many of the program topics and helped plan the sessions. FOREST INDUSTRIES and WESTERN TIMBER INDUSTRY, sponsors of the Clinic, and MILLER FREEMAN PUBLICATIONS, INC. thank these Advisory Committee members:

James W. "Jim" Bigelow, President, Bigelow Machinery Inc., Portland, Oregon

Jack Gates, President, 3-G Lumber Co., Philomath, Oregon

John Gram, Vice President, Gram Development Co., Portland, Oregon

Paul R. Hollenbeck, Executive Vice President, West Coast Lumber Inspection Bureau, Portland, Oregon

A. L. Hurl, Partner, Oregon Alder-Maple Co., Willamina, Oregon

Everett P. "Bud" Johnson, Partner, C & D Lumber Co., Riddle, Oregon

Aaron U. Jones, President, Seneca Sawmill Co., Eugene, Oregon

Paul Kornberg, West Coast Manager, Soderhamn Machine Manufacturing Co., Portland, Oregon

Arthur Lindley, General Manager, Kimball Bros. Lumber Co., Dexter, Oregon

Lloyd Porter, U.S. Department of Commerce, Portland, Oregon

Roy Sage, Sales Manager, Sherman Machine Works, Woodland, Washington

Charles E. Stanford, Vice President, Portland Machinery Co., Portland, Oregon

Bud Stewart, Trainee, Bohemia, Inc., Eugene, Oregon

Ray Swanson, President, Swanson Bros. Lumber Co., Inc., Noti, Oregon

Speakers at the first 1973 Sawmill Clinic had an attentive audience.

Chapter Summaries & Clinic Authors

CHAPTER 1   **SMALL LOG HANDLING &**
SORTING SYSTEMS
By James H. Porter

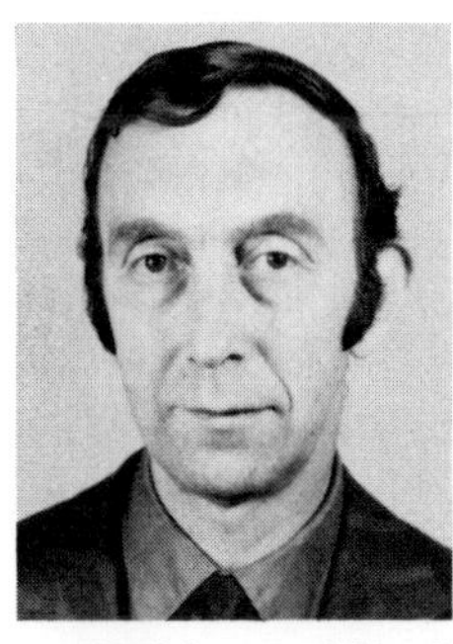

JAMES H. PORTER, chief engineer for CAE Machinery, Ltd., Vancouver, British Columbia, joined that firm (then Canadian Sumner Iron Works) in 1962, starting as a design engineer and moving through chief draftsman to his present post. Educated in England, he holds a mechanical engineering degree from Lincoln Technical College and is licensed in British Columbia. He has worked in civil engineering and surveying in British Columbia and was surveyor for a mineral prospecting team in Sierra Leone. Porter has studied log and lumber sorting systems in Scandinavia and has traveled there, in Europe and in Australia and New Zealand in connection with wood handling and processing machinery. He participated in the wood processing symposiums sponsored by the Canadian and Soviet governments at Archangel, USSR, in 1972 and has worked closely with Hammars of Sweden to adapt its log handling system for use in North America.

CHAPTER SUMMARY: This chapter traces the history of log sorting systems from their beginnings in Scandinavia about 1957. Sorting to size permits close flow and eliminates the need to reset heads or saws between logs. To produce a face smooth enough for acceptance in a market for unplaned lumber, Scandinavian mills operate at slower speeds than do North American mills.

Hammars of Sweden has built some 50 sorting systems; CAE Machinery Ltd. has cooperated in developing nine for North American use.

Bucking is usually integral with sorting in North America. A sorting system needs high capacity coupled with superior control; a maverick can jam any system. Basic to the success of a sorting system is the support of those who will operate it.

An automatic system has five components: (1) infeed, (2) sorting conveyor, (3) outfeed, (4) cleanup system and (5) controls. A station-by-station walk through a model sorting system is described; alternatives in use at various plants are noted. Reasons are given for the choice of air rather than electric or hydraulic controls.

CHAPTER 2   SATEKO LOG SORTING SYSTEM

By Robert E. Vadnais & Dick Johnson

ROBERT E. VADNAIS, president of Totem Equipment Co., Seattle, Washington, was born in Portland, Oregon, in 1917. He attended Jefferson High School in Portland and the University of Oregon in Eugene. In World War II, he spent six years in the Navy Air Corps, seeing action at Okinawa, Iwo Jima and in the Philippines. After the war, he became manager of Nelson Equipment Co.'s Seattle branch, which he bought in 1954 to form Totem Equipment Co. Totem represents Pacific Car and Foundry Co. in five states for Dart log stackers and front-end loaders. More recently, Totem has added the Sateko log sorting system to its lines.

DICK JOHNSON, construction and maintenance superintendent for Van-Evan Co., Missoula, Montana, was born in the Rocky Mountain region at Sheridan, Wyoming. He attended Montana State University in Missoula, emphasizing courses in the physical sciences. During college, he began working part time at the Van-Evan plant. He moved through master mechanic and plant engineer positions with Van-Evan to assume his present post in 1970. As construction and maintenance superintendent, he ramrods and troubleshoots plant additions and new installations and sees that the plant is maintained in smooth working order. [Photograph not available.]

CHAPTER SUMMARY: Before sorters were developed, mill operators often used a $120,000 stacker to sort logs. Then they tried to buck on the infeed deck, but costs were prohibitive, Vadnais, adapter of the Finnish Sateko sorting system for North American use, points out. The first system has been installed at the Van-Evan mill at Missoula, Montana.

The Missoula installation has a three-section infeed deck, a crossfeed conveyor and a surge deck. Log centers are determined electronically for stable log seating on the conveyor. Erection of a model to test electronic centering has eliminated need for a crossfeed conveyor and surge deck, cutting costs $60,000 and eliminating one man, unless bucking is desired. Vadnais believes sawing on the infeed deck will be permitted later.

Operational problems, particularly with jackstrawing and breakage of logs entering bins, are enumerated and a solution using hog fuel bedding is described by Vadnais.

Johnson reports that the modified Sateko sorting system has worked well with the large variation in log diameters (4 to 40 inches) at the Missoula installation. In the Van-Evan operation in Missoula, the selector is geared to the idler sprocket on the end of the sort chain. Relationship is constant between sprockets and a roll of bars. When a key is pushed when the log rolls onto the sort conveyor, the log has already been centered on the chairs. The operator keypunches species and diameter class. There are 15 rows of bars, 24 degrees apart on a circle. There are two rows of microswitches, 11 in each. Pressing a key causes a cam action. As the cam travels, it hits one of the 22 microswitches, which pulse an electromagnet; the magnet trips a finger, selecting the right pocket.

CHAPTER 3 ELECTRONIC LOG INVENTORY CONTROL SYSTEMS

By Ed McGrew

ED McGREW of McGrew Bros. Sawmill Inc., Ashland, Oregon, attended Oregon State University and graduated from Southern Oregon College in 1966 in business management. While attending college, he worked as a chaser on logging landings, as powder monkey on road crews and, in 1962, gave up setting chokers to puddle jump on the mill pond. He was tail sawyer, planer feeder and cleanup—and drove a log stacker. After college, he spent a year with Standard Oil at Coos Bay, Oregon, then joined McGrew Bros. as purchasing agent and safety director. Ed McGrew is a member of the third generation of McGrews in the sawmill/logging industry in southern Oregon: His grandfather ran a water-powered sawmill; his father, "Pink" McGrew, still active, worked with horses and oxen and, with his brothers Melvin and Curtis, formed McGrew Bros. Logging Co. In 1952 the McGrews bought a sawmill in Medford, Oregon, and in 1963 acquired a planing mill at Ashland, Oregon, and built a sawmill there. The mill produces about 50,000 board feet of lumber yearly and employs 165 people. Ed McGrew began working on log inventory control systems in 1968, installing the first developed system in 1969 at the McGrew operation in Ashland.

CHAPTER SUMMARY: The Log Inventory Control System (LOGINV) grew out of a search that began in 1968 for a better management information system for McGrew Bros. at their Ashland, Oregon sawmill. The tag system of log inventory control led to cumulative error—wrong data were keypunched, tags were separated from logs, logs were broken and so forth. Manual inventories have an error factor of up to 10 percent.

Six reports, one of them on log inventory, were developed on a time-shared computer with Computer Science Corp. No information system is better than the data fed into it. Conventional scaling and optical scanning contain error and results often are delayed up to 10 days.

To reduce or eliminate errors in data, an electronic device built by Entron Inc. permits the scaler to key information, display it for accuracy, then add it to the Dataholder's memory. The Dataholder is rugged; it includes length, diameter, grade, species and defect. The operator dumps the load information into a "black box" or formatter, which directs the teletype to print a scale ticket on hard copy and on paper tape or magnetic tape. At shift end, the paper tape or magnetic tape cassette goes to the office for transmittal to the computer via telephone line.

A man at the barker makes similar calculations at the mill infeed. The machine automatically deducts an average defect, making the deduct side of the Current Inventory Report. Similar reports on trucking, timber sources and daily and period mill usage have been prepared.

McGrew estimates cost of the system at about 4½ cents a log for mills in the 30- to 40-million-board-foot category.

CHAPTER 4   BEST OPENING FACE FOR SECOND-GROWTH TIMBER

By Hiram Hallock

HIRAM HALLOCK, technologist in the Sawmill section of the Division of Wood Quality Research, U.S. Forest Products Laboratory, Madison, Wisconsin, graduated from the University of Minnesota in forestry in 1942. After graduation, Hallock moved to Crossett, Arkansas, as a district manager in forestry and logging for Crossett Lumber Co. From 1945 through 1957, he operated two sawmills in southern Wisconsin, adding a pressure treating plant and spending considerable time on mill equipment development. He then spent one and one-half years as a sawmill consultant to the government of Iran. Since joining the U.S. Forest Products Laboratory in 1959, Hallock has studied the pulpability of sawdust, kerf and the relation of sawing pattern to warp in pines. Presently he is working on the Best Opening Face (B.O.F.) and Edge, Glue and Rip (EGAR) programs at the Madison, Wisconsin laboratory and on a decision-making program for edging, trimming and upgrading hardwoods.

DAVID W. LEWIS, a technologist in the Structure-Properties section of the Division of Wood Quality Research, U.S. Forest Products Laboratory, Madison, Wisconsin, did not participate in the presentation of this paper at the 1973 Portland Sawmill Clinic but he did collaborate and consult with Hiram Hallock on preparation of technical data. Lewis also was present to answer technical questions on the data base and the computer systems and programs involved. Lewis received his bachelor and masters degrees in forestry from the Universities of Maine (1961) and Wisconsin (1969) and has spent a year of full-time work on computer science. Lewis developed the computer program for Best Opening Face (B.O.F.) and also designed the data program for the Edge, Glue and Rip (EGAR) program. He has studied the relation of structure to specific gravity and compressive strength of normal and tension work, the effects of gamma radiation exposure and similar fundamental questions. Electronic detection of defect and optimum decision-making in the lumber mill on this problem presently occupy his time.

CHAPTER SUMMARY: The Best Opening Face (B.O.F.) program of the U.S. Forest Products Laboratory is a model of the sawing process, aimed at maximizing yield. Location of the opening face on log or cant is the key to the sawing program that will yield the most lumber. Shifting the saw line across the log end results, in most cases, in substantial differences in potential lumber yield. Small changes in cutting pattern affect yield more in small logs than in large.

Diagrams of the sawing process can examine only a fraction of possible sawing combinations. They are worthless if one factor—such as kerf—is changed. B.O.F. was designed to examine all sawing solutions. It considers dry finished lumber size, planing allowance, shrinkage, sawing variation, kerf width, log diameter range, diameter measuring increment and equipment's capability to produce a given opening face.

The best live log sawing solutions were found to be from 6 to 90 percent better than the poorest—averaging 21 percent better. The best cant sawing solutions exceeded the worst by 12 to 100 percent and averaged 27 percent better. Since the fitting problem is reduced, differences tend to decline with larger diameter logs.

Lumber recovery in North America ranges from less than 4 to more than 9 board feet per cubic foot of log; the average is about 6½. Optimizing the sawing geometry promises major gains in lumber recovery.

Newer refinements in the B.O.F. program take into consideration (1) taper, (2) width of resaw kerf, (3) narrowest and shortest acceptable lumber and (4) wane. The system "dollar maximizes"—considering price relationships of all lumber items—as well as maximizes yield.

The program finds solutions producing higher recovery rates than the industry average for edging and trimming.

It is estimated that the B.O.F. system can be completely installed in a new mill for approximately $125,000. Payout time for a mill cutting 50,000 board feet per day would be about one year.

CHAPTER 5   CHIPPING HEADRIGS & PROCESS REQUIREMENTS

By James L. Gregoire & Ernest Helvogt, Jr.

JAMES L. GREGOIRE, general manager of Chip-N-Saw, Eugene, Oregon, was born in Grandview, Wisconsin, and attended the University of Minnesota before entering the U.S. Army, where he served under General Patton in western and central Europe. He began his sawmill experience as a filer with Top Hat Lumber Co. near Redding, California, opening his own saw filing shop and adding sales lines in the filing field. At the same time, he spent seven years in the lumber business with his father in the Redding area, then bought a gang mill to help with the cut from the Trinity Dam reservoir area. Gregoire moved to Irvington Machine Works in Portland, then to Standard Dry Kiln, Indianapolis, Indiana, as a sales manager. Standard sent him back to Portland to manage West Coast operations. In Portland, he rejoined Irvington for four years as sales manager. When Canadian Car cancelled dealerships and installed its own sales office, Gregoire continued to handle promotion of Chip-N-Saw as general manager of Can Car's Chip-N-Saw division. The office was moved from Portland to Eugene, Oregon, in the spring of 1973. Gregoire hunts and pilots his own plane. He has traveled extensively in Latin America for Irvington and Standard.

ERNEST HELVOGT, JR., a division engineer with Crown Zellerbach Corp., Portland, Oregon, for the past six and one-half years, graduated from Oregon State University in industrial administration in 1951. A Portland native, he began his career as a sales engineer with Alaskan Copper & Brass, then moved to Container Corp. for 10 years as a plant engineer, working with equipment layout, design and installation. He managed the printing division at Kenton Machine Co., working indirectly with plywood stackers and presses. He was chief engineer at Timber Structures, working on mill design and material flows, and was also with Alta Engineers before joining Crown Zellerbach. Helvogt spent four years in the U.S. Navy, part of it in the Naval Air Corps.

CHAPTER SUMMARY: The development of chipping headrigs is traced by Gregoire—from the first Chip-N-Saw installed in 1963 to the more than 250 now in operation on three continents and New Zealand. Such canters can accept logs to 28 inches in diameter and can work with twin and quad band resaws. Small machines take peeler cores. There are chipping edgers. All add to usable fiber by eliminating outside kerf. Half-taper configuration gives best recovery in logs from 5 to 11 inches while full-taper provides best yield in larger logs. Potential gains in usable fiber from various log sizes are analyzed in relation to chipping headrigs and edgers. Recovery with a chipping headrig and with traditional sawing are compared in terms of board feet of lumber, amount of chips and sawdust produced. On a typical 9-inch-diameter log 16 feet long with 1/16-inch taper per foot, the chipping headrig produced 13.87 percent less sawdust through elimination of the outside saw kerf line.

Helvogt describes the Crown Zellerbach mill at Columbia City, Oregon, where a Chip-N-Saw Mark II is in operation. The machine cants 6- to 14-inch logs, 95 percent hemlock, at 120 feet per minute. The Chip-N-Saw quad has run two shifts per day with little maintenance since installation in 1968. The mill recovers about 750 board feet lumber per cunit log input. Taper runs full length. Analyses of recovery measurements based on cubic volume (logs in and lumber out) and on overrun are presented and shortcomings in measurements based on overrun are pointed out. A scanner, which will monitor the Chip-N-Saw installation has been installed at the Columbia City mill. Helvogt notes that chipping equally from both sides relieves stress, resulting in straighter pieces with less twist and sweep; propellering is practically eliminated. When using a quad or a twin behind the chipping section to cut boards, thin saws can be used to a great extent. Saw kerf at Crown Zellerbach's Columbia City mill is 1/8 inch.

CHAPTER 6 DOUBLE-TAPER CHIPPER CANTERS

By Gilbert W. Anderson & John L. Ailport

GILBERT W. ANDERSON is president of Stetson-Ross Machine Co. Inc., Seattle, Washington. A native of Seattle, he graduated from the University of Washington in economics in 1952. He served in both World War II and the Korean conflict. Before being named president of Stetson-Ross in April 1969, he was vice president and general manager of United Control Corp., a manufacturer of electrical control systems; he was with United Control for 13 years. He is a member of the safety standards committee of Woodworking Machinery Manufacturers of America (WMMA). Anderson is particularly interested in the development of foreign markets for wood processing machinery, an interest that has taken him to the Orient, Scandinavia and Europe in support of foreign sales and licensing.

JOHN L. AILPORT, plant manager, Diamond International, Albeni Falls, Idaho, was born in the sawmill town of Bonner, Montana. He graduated from Missoula High School and enrolled at the University of Montana's School of Forestry. He served as a Forest Service fire lookout and worked at carloading, moulding grading and similar jobs at The Anaconda Co.'s (now Champion International's) Bonner sawmill, then as a faller and log truck driver. He returned to the mill as planer machine feeder and setup man until joining the U.S. Navy in April 1942. While in the Navy, he completed machinist school and was assigned to a destroyer engineroom for 26½ months until the ship was sunk in the Philippine campaign. After receiving his discharge in September 1945, he returned to the University of Montana forestry school, receiving his degree in 1952. While attending school, he worked as a blacksmith and also helped in construction of two sawmills and in log truck repair. After college, Ailport returned to Anaconda as dry kiln foreman, moving to planer foreman, plant superintendent, sales manager and production manager positions. He laid out the planing mill, laminating plant and the sawmill remodeling at Anaconda. In 1964, he joined Diamond International as plant manager at its Albeni Falls, Idaho mill and was recently promoted to his present position. He has had charge of planning and construction of the new sawmill and planer installations at Albeni Falls. Ailport belongs to the Forest Products Research Society and the Society of American Foresters. He is past president of the Northwest Wood Products Clinic. A Chamber of Commerce director and past president of the Western Montana Fish and Game Association, Ailport also holds the Silver Beaver Scouting award and serves on the Boy Scouts' western Montana council.

CHAPTER SUMMARY: The double-taper chipper canter reportedly recovers more lumber from a log than conventional machinery through its handling of taper. Anderson traces the machine's development, from early discussions with Diamond International. Diamond's objective was up to 150,000 board feet per shift from logs under 15 inches in diameter. Stetson-Ross developed the double-taper chipper canter to chip four parallel faces. Fixed bottom and side heads chip full-length faces. Additional lumber is recovered from the taper accumulated along two axes. Adjustable chipperheads chip top and opposite side faces if taper is sufficient. Taper is oriented to affect only two boards, minimizing production of shorts. Cants are fed to fixed linebar twin resaw. System requires only one man. Chips are reported excellent.

Ailport traces the history of attempts to maximize recovery through various sawing approaches to log taper. Half-tapering was traditional until mills became concerned about maximum utilization. Then several methods, each with a different effect upon grade recovery, were developed. The double-taper method permits recovery of 6- to 12-foot boards on the side opposite the double taper. Diamond International Corp. uses a scanner at its Albeni Falls, Idaho mill to determine the best cutting method. This in-line method speeds production. Cants are more accurate, as the log is held against a solid bedplate. Multiple bands permit slower speeds, improving accuracy. Accuracy plus thinner kerf equals greater yield.

CHAPTER 7 — ADVANTAGES OF HEADRIG CHIPPERS
By Fred Miller & Frank Roppel

FRED MILLER, chief engineer for Chipper Machines & Engineering at Lake Oswego, Oregon, graduated from the University of Idaho in 1935 with an M.S. degree in mechanical engineering. He served as a lieutenant commander in the U.S. Navy during World War II and was stationed in the South Pacific as an engineer on ship construction and repair. Miller has invested 35 years in development and design of heavy equipment, 20 of these working on equipment for wood waste utilization. He has been heavily involved with chipping machinery since his firm pioneered the introduction of chipping headrigs, based on the old Standal patents. Miller holds a number of patents in the woodworking industry.

FRANK ROPPEL, plant manager since 1971 for Ketchikan Spruce Mills' operations at Ketchikan and nearby Metlakatla, Alaska, received a degree in business administration from Oregon State University in 1959. His family had been involved in the pulp and paper industry since the 1930s, primarily with Crown Zellerbach Corp. Ketchikan Spruce is jointly owned by the Louisiana-Pacific and F.M.C. corporations and Roppel was plant manager for the Willits, California operations before the L-P spinoff from Georgia-Pacific Corp. Previously he had been in accounting with Ketchikan Pulp Co. for eight years and had been appointed chief accountant for Ketchikan Spruce in 1968. Prior to his appointment as a Ketchikan Spruce plant manager, he spent part of 1970 as a management trainee. Roppel has specialized in marketing sawn lumber from southeast Alaska and has spent time in Japan in connection with this activity.

CHAPTER SUMMARY: Single-cut headrigs report 15 to 30 percent production increases with Chipper Machines & Engineering Corp.'s headrig chippers; installation cost is $60,000 to $80,000, Miller reports. The double-cuts, when supplemented with a headrig chipper, will generally show about a 15 percent production rise. Advantages of the headrig chippers include conversion of kerf to chips; elimination of offbearer in some instances; removal of flares, burls and crook before they cause jams. Material previously hogged, burned or made into residue chips becomes quality chips. One mill reported a 12 percent chip production increase. Chip quality rises because the log is stable, whereas chunks in a wastewood chipper are not. Sawyer attention need no longer be diverted to offbearer's safety. Sawyers prefer position over the slabber; some have been able to retrieve previous production levels within one shift. Chipper edgers with the addition of a top head can reach feed rates up to 600 feet per minute.

Roppel observes that Ketchikan Spruce Mills has saved two lines (out of an average of five) with the CM & E headrig chipper. The slabber has increased production an estimated 25 percent. The offbearer's work load has been reduced. Cut quality is acceptable for the Japan export market. Knife change on the headrig slabber takes about 15 minutes with socket wrench. The filer maintains the slabber. Reduction in the amount of sawdust is noticeable and there is likewise an increase in chip recovery. Slabs previously too wide for the chipper throat have been diverted from burner to chips.

Knives will break if improperly tightened without cleaning the holder; they will also break if not properly ground and released in the radius of the curve. Knives are ground in the filing room. Grinding takes approximately five minutes per knife on a single-holder knife grinder.

Using shadow lines with the slabber to prevent overfeeding is recommended. Cut depth limit is 6 inches. Chip quality in frozen logs is not as high as in unfrozen logs.

CHAPTER 8   INCREASED PRODUCTIVITY WITH HEADRIG SLABBERS

By Jim Bigelow & Cliff Jackson

JIM BIGELOW, western U.S. dealer for Beloit-Passavant's sawmill machinery (including the slabbing chipper, barkers and other chippers), was born in Seattle, Washington. He completed his education there after spending part of his youth in New York and Connecticut. He served in the Pacific with the U.S. Navy during World War II. He was closely involved in the development of portable spars for the logging industry. Long a dealer for the LeTourneau log stacker, Bigelow was a pioneer in introducing dry log handling on the West Coast and is familiar with Scandinavian sawmill practices. A devoted yachtsman, he has also completed a Baja roadrace through lower California.

CLIFF JACKSON, plant engineer and maintenance superintendent for American Forest Products Corp.'s (a Bendix Corp.) Amador-Calaveras division, Martell, California, holds two degrees in engineering. Formerly chief engineer for one of the world's largest sawmill manufacturers, Jackson has spent a lifetime in sawmilling and related industries. He had the unique experience of singlehandedly subduing one of the early skyjackers after the plane's co-pilot had been shot—and was credited with saving the flight. He is the holder of Boy Scouting's high award, the Silver Beaver. An enthusiastic hunter and fisherman, Jackson is a frequent visitor to the Pacific Northwest hunting and fishing grounds.

CHAPTER SUMMARY: Bigelow reports that the Beloit Passavant headrig slabber increases production up to three lines per log. It eliminates swelled butts and can often eliminate the offbearer function. The slabber makes thinner edger saws possible by chipping swelled butts. Chips are whole-log chips, not wastewood chips. A soundproofed, air-conditioned cab aids sawyer visibility and efficiency and reduces sound decibels below legal limit. Sawyer preparation and training prior to slabber installation is discussed. Weekend installation eliminates bandsaw downtime.

Assets of the Beloit-Passavant slabber include: (1) one-piece, straight-knife design, (2) knives are available from most suppliers and can be ground on ordinary mill equipment, (3) decrease in suck saves air on carriage dogs, (4) narrow design fits existing saw boxes and (5) equipment is rugged.

American Forest Products Corp.'s concerns in selecting a slabber for its Martell, California plant included narrowness of available space and budget, Jackson reports. The Beloit-Passavant slabber was chosen. The sawyer's cab was suspended from the filing room floor, to avoid vibration from the saw. It is suggested that nigger cylinders be located farther from the saw line to prevent logs turning in front of the slabber head, that a chain (rather than a belt) chip conveyor be installed and that a flop gate be provided to divert chips to the hog if the chip line malfunctions.

The slabber at American Forest Products was installed on weekends. A single console contains all controls. Television camera behind the bandmill gives sawyer a view of the bumpers to the resaw and edger. Knife grinding problem was corrected with a simple gauge. Jointer can smooth the slab finish; a modified planing mill jointer is used at the Martell mill.

CHAPTER 9	**HIGH-STRAIN/THIN KERF** **By F. E. "Ed" Allen**

F. E. "ED" ALLEN, chief engineer, Letson & Burpee, Ltd., has been with that company since his discharge from the Royal Canadian Electrical & Mechanical Engineer Corps at the close of World War II. He has held his present position for 12 years, having served previously in projects manager and other engineering posts except for a period (1953-1955) in charge of production control. Allen was design draftsman in development of the "Tormag" permanent magnetic drive. Just prior to attending the 1973 Sawmill Clinic in Portland, Allen returned from a lecture tour of Australia and New Zealand where he was guest speaker at the annual conference of the Forest Industries Engineering Association. His paper, "Quality Control in the Lumber Industry," will be published in the *Australian Forest Industries Journal. Forest Industries Review* has published his papers on bandsaw techniques. A member of the Canadian Society for Mechanical Engineering, Allen believes his principal contribution to the industry has been to replace black magic with science in the bandmill field.

CHAPTER SUMMARY: The failure of a Quebec mill to remove packing blocks around the upper wheel of a bandmill gave birth to the theory of high-strain/thin kerf. Work was in frozen logs. When strain was reduced, the saw snaked. This fact, coupled with information on metal fatigue, formed the basis for bandmill design and for determining safe working stresses with strain applied.

Fluctuating stress (saw revolving around the band wheel, then straightening) causes fatigue failures. To reduce fluctuating stress, a thinner saw is needed. Practical limit is 1/1500 of wheel diameter. Correct gauging and roll-tensioning overcome anticlastic curvature.

High-strain is a balance of stress and strength. Prime purpose of a bandsaw strain system is to protect the bandsaw from damage. The strain system must respond instantly and effectively to any happening in the cut. Letson & Burpee has run tests to measure strain system response; results are reported and the influence of strain on accuracy is discussed.

CHAPTER 10	**TRADEOFFS AT THE HEADRIG: PRODUCTION VS. RECOVERY** **By P. S. "Phil" Quelch**

PHIL QUELCH, author of *Sawmill Feeds and Speeds,* was born in Manitoba, Canada and entered his first filing room in 1924 at Ladner. He worked at several sawmills in British Columbia and Alberta until 1930, when he became head filer for Pacific Mills. He moved to the same position at MacMillan Bloedel's Chemainus mill from 1935 through 1951. He was then named western manager of Simonds Canadian Saw Co. After his retirement in 1969 from Simonds, he began his consulting career. He has trained bandsaw filers in Sweden and instructs at British Columbia's Institute of Technology. Before writing *Feeds and Speeds,* he authored *The Filer's Handbook.* Quelch lives in West Vancouver, British Columbia. As a continuing hobby since 1939, he has built two sawmills, scaled one inch to one foot. They are the world's smallest operating mills and have been exhibited in major cities throughout both American continents.

CHAPTER SUMMARY: The capacity of a bandsaw should be known and its speed and feed balanced. Early efforts to determine a saw's capacity are traced. The author maintains that both U.S. and Canadian Pacific Northwest mills overbite the sawtooth and gullet capacity; mills in other areas generally underbite. Minimum bite is the clearance on one side of the saw plate. Smaller bits spill from the gullet, pushing the saw off the saw line. Collecting sawdust in the gullet and discharging it to free air is the problem in sawing— cutting wood is not. Compressing sawdust beyond gullet capacity can create pressure on the gullet edge between 1,600 and 2,000 pounds per square inch.

Heavy feeds require heavy saws. If saws were not fed beyond rated capacity, thinner saws could be used and accuracy maintained. Too often, saws are overfed.

It is estimated that solid wood equal to 70 percent of the gullet area can be capably conveyed. In deep cuts, widening tooth space to add gullet is recommended. Development of an indicator displaying feed rate and a load meter indicating power consumption would provide valuable and necessary information to the sawyer.

CHAPTER 11 **OPERATIONAL EXPERIENCE IN AUTOMATED SAW CONTROL**
By Dan Pichulo & Daniel F. Williamson

DAN PICHULO heads marketing management at Atmospheric Sciences Inc. (ASI), Sunnyvale, California. He graduated from Tufts University in Massachusetts in 1954 with a Bachelor of Science degree in chemical engineering. After graduation, Pichulo joined Du Pont's Atomic Energy Division in Augusta, Georgia, as process construction engineer. He remained with Du Pont until 1969, ultimately becoming manager for systems marketing and sales at the company's Wilmington, Delaware, headquarters. Pichulo has worked with the Dale Carnegie leadership courses as an associate instructor. He has taken a number of American Management Association courses in marketing, finance and systems analysis. In 1969, Pichulo moved to Los Altos, California, forming his own company in motivation management. He was president there until April 1972, when he joined ASI. Pichulo's responsibilities at ASI include coordination of marketing and sales activities for the southern United States. His background includes marketing and systems experience for industries such as electronics, refrigeration, photographic film, primarily in production operations and product development. Most recently at ASI, Pichulo has been active in development and marketing of electronic measurement and computerized process control systems now coming into use in sawmills. ASI has delivered its system for more than 60 lines in sawmills and plywood plants throughout North America.

Pichulo is a strong advocate of computerized process control systems for production of information to assist management in making better and more comprehensive decisions.

DANIEL F. WILLIAMSON, executive vice president of Publishers Paper Co., has charge of day-to-day operation of the company's eight pulp and paper, lumber, plywood and hardboard mills in California, Oregon and Washington. He was a Flying Officer in the Royal Canadian Air Force during World War II and graduated from the University of British Columbia in civil engineering in 1949. He joined Columbia Cellulose Co. Ltd. as a field engineer during construction of the 300 ton-per-day bleached sulphite dissolving mill at Prince Rupert, British Columbia. In 1951, Williamson joined Sandwell & Co., Ltd., gaining design, project and engineering management experience, which led to his appointment as vice president. He moved to Portland in 1963 as chief engineer of Sandwell International, Inc., later becoming vice president and manager of the firm. In 1970, he was appointed president of Sandwell International. He joined Publishers Paper in April 1971 as executive vice president and was elected a director of Publishers in June 1972.

CHAPTER SUMMARY: The key to increasing yield is knowing where to cut the log; this requires accuracy in measurement. Often 40 percent of a system's sets require the sawyer to discern the difference between ¼ inch and ½ inch; 75 percent require differentiation between ¼ and ¾ inch. Sawyers cannot make the decisions required accurately on logs moving 100 feet per minute for eight-hour shifts. Optical technology with solid-state circuitry has permitted the development of new types of scanners. They are compact and can be mounted remote from vibration. With length, diameter and profile scanned, the best saw program is set instantly and digitally displayed. The sawyer can override. ASI scanners will fit any equipment in any mill.

A major benefit of electronic measurement and computerized process control is the management report which makes mill management's decisions more comprehensive and effective. This information becomes available because measurement has already taken place. Managers are inundated with information; trying to bring order out of it is difficult. With computerized control, the management report is available instantly and can be produced at any hour. A manager can see a trend developing, since the report is really a feedback report. If a manager makes a change in operating procedure or adjusts equipment, he can track the effects of changes immediately.

Williamson believes that efficiency can be raised by reducing decisions for the sawyer. The key decision is in the first cut at the log. Publishers Paper Co. runs two quads at its Molalla, Oregon mill, feeding three to five logs per minute—seven in perfect conditions. Sawyers cannot make correct decisions at such speeds for extended periods. Computers can.

About 95 percent of Publishers' Molalla mill logs are sawn with automatic sets. There are 2,000 cutting patterns at Molalla and 2,800 at Burney, California. Investment should be repaid within the first year. At one mill, a sorter tally report is being added. A log bucking program to maximize market conditions is underway. The variation of market prices will also be plugged into the system.

Production and maintenance people should be included in equipment decisions; failure to do so delayed installation of the scanner at the Molalla mill.

CHAPTER 12   COMPUTERS IN THE MILL ENVIRONMENT
By Dr. Don Anderson

DR. DON ANDERSON has been director of computer software development programs at Black Clawson, Inc., Everett, Washington, since 1972. He was born in Bremerton, Washington, in 1938. Dr. Anderson's extensive academic background uniquely prepared him for computer software work in the forest products industries. He holds bachelor degrees from Central Washington College (in mathematics) and Reed College, Portland, Oregon, (in history) as well as a doctorate in physical chemistry from Washington

State University. More recently, he has done postdoctoral work at the University of North Carolina on recognition of patterns by computers. Anderson writes programs for mill automation and application, with particular emphasis on systems and programs for detection of log sweep and wane edge on cants and pattern recognition.

CHAPTER SUMMARY: More accurate scaling information is making it possible to determine whether slight changes in machinery actually improve recovery; this could not be done in the past. Recovery improvement lies in wise use of digital computers coupled with scanners. These give sawmill personnel the tools to do the job more effectively.

General purpose digital computers contain (1) a memory, which stores both data about the log and the cutting programs, (2) a central processing unit, which executes instructions and (3) connections with the outside world. Computers have become more rugged but will still shut off with irregularities in voltage.

Studies are reported showing losses of 9 percent of potential recovery at the edger—2,624 board feet or $255 per shift at one mill. Based on these figures, a scanner/computer combination with setworks would eliminate the loss and pay for itself within a year.

The author predicts that small computers at each mill work station—rather than a huge centralized system—will ultimately prevail.

CHAPTER 13   **RECOVERY AT THE RESAW: BAND VS. GANG**
By Jack Gates

JACK GATES, president and owner of 3-G Lumber Co. near Philomath, Oregon, first worked in the woods at age 13. Born in Corvallis, Oregon, he graduated from Oregon State University in forestry and logging engineering. He spent five years in engineering with Boeing Co., Seattle, in charge of such mechanical items as airplane doors and latches. Gates then returned to Corvallis and established his own logging firm. He bought a steam-powered, nonautomated mill in 1949 and has completely modernized it; none of the mill's original equipment remains. Gates is very active in industry affairs, serving in 1973 as vice president both of Western Forest Industries Association and of the West Coast Lumber Inspection Bureau.

CHAPTER SUMMARY: Advantages of a band resaw over a cant gang for large second-growth logs with coarse grain and large knots (the timber available to 3-G Lumber Co.) are presented. The author believes that more mills must cut this type of log in the future.

Logs are sawn to square cants, centering pith. All knots dimension to zero diameter at the pith. First cuts produce 2x12s with largest knots. Each successive cut is taken from the side with largest knots. Lumber grading rules permit larger knots in wider surfaces. As board size decreases and cuts advance toward the center, knot size also decreases, resulting in the least grade loss for oversized knots. Spike knots are eliminated. In contrast, a gang makes all cuts at once, precluding rotation.

Band resaws permit changes in sawing pattern when defects appear on the surface. Outflow from a band resaw is easier to handle than multiple-piece flow from a gangsaw. Bands produce a smoother lumber surface, as gangs tear wood when slowed at the top of the stroke. Resaws are extremely versatile in reducing cants to any desired dimension size without loss in board foot volume.

CHAPTER 14	SELECTING, SAWING & SELLING SPECIALTY ITEMS

CHAPTER 14 SELECTING, SAWING & SELLING SPECIALTY ITEMS

By Ray Swanson

RAY SWANSON is president of Swanson Bros. Lumber Co., Noti, Oregon. He became manager of the mill, founded by his father, H. R., and uncle, Walter Swanson, in 1948 and has been president since 1968. He has worked in the lumber business nearly 40 years, starting in the old Coast Basket & Veneer Co.'s box shook plant. He also worked in plywood before becoming head sawyer at Swanson Bros. Ray's brothers, Rod and Dean, are principals in Superior Lumber Co. at Glendale, Oregon. Swanson is active in industry affairs, having served as president of Western Forest Industries Association for three terms; he is a member of the board of trustees of the West Coast Lumber Inspection Bureau. He served 20 years in a lay capacity in public education, culminating in a four and one-half-year term on the Oregon State Board of Education. In 1970 he was the Democratic nominee for the Oregon State Senate from Lane County.

CHAPTER SUMMARY: So many mills specialize in scant-sawn standardized sizes that items which used to be common—long timbers, stringers, ceiling and siding patterns and millwork—now are regarded as specialty items.

Manufacture of lumber is a separation process. Except for gemstones, nearly every other manufacturing process adds value by combining elements, not by separating them. Once separated, lumber's elements cannot be recombined for a new try. Consequently, great care is required in the lumber manufacturing process.

Woods and mill crews working with specialties must be proficient and have good attitude to produce quality workmanship. Mismanufactured dimension loses only grade; fractional-sized material also loses scale. Therefore, production people need to understand product end use. The importance of decisions made by the timber faller and the bucker on the structural and economic value of the final product are stressed. Sawyer decisions on taper, wane, cut depth and location of defect are discussed.

Markets for established specialty items from old-growth are dying faster than new ones are opening. And dwindling supply makes users seek substitute materials.

<table>
<tr><td>

CHAPTER 15

</td><td>

TRIMMER CONTROL &
LUMBER TALLY SYSTEMS
By Edward N. Rickford & Gene Neely

</td></tr>
</table>

EDWARD N. RICKFORD, president of North American Controls, Inc., Portland, Oregon, graduated from the University of Edinburgh, Scotland, in 1959. He worked in German forests and sawmills before returning to his native Guiana, South America. Rickford then spent seven years as assistant conservator of forests in state-owned tropical hardwoods. He organized an advisory service to the tropical hardwoods sawmill industry, with designs based on European equipment. In 1967 he emigrated to the United States to become consultant to Northwest Hardwoods, Inc., manufacturers in alder and maple. He spent four and one-half years as a project manager with H. C. Mason & Associates Inc., Gladstone, Oregon, studying economic feasibility and justification for capital investments in new sawmill design and equipment. He formed North American Controls in the first half of 1973 to "take theory into the field."

GENE NEELY, executive vice president of North American Controls, Inc., graduated magna cum laude from the University of Montana in forestry in 1965 and has since earned a master's degree in business administration. He spent four years in the U.S. Air Force, where he supervised maintenance and design on computer-based aircraft systems. He has been with the University of Montana's Business and Economic Research Bureau and with the Montana Forest Conservation and Experiment station as a wood technologist. Before joining American Controls, he was a project engineer with H. C. Mason & Associates Inc., sawmill engineers. Neely has developed optimum sawing programs for computer control of various log breakdown equipment and the report systems that can be obtained. He has also designed and developed computer-based systems to control grade clipping of veneer and to optimize lumber recovery with Chip-N-Saws.

CHAPTER SUMMARY: High-strain/thin kerf bandmills and high-precision, thin kerf edgers have increased recovery potential. Computerization now makes it possible to predetermine in theory best sawing sequences by individual log sizes down to 1/10-inch-diameter increments. Rickford reviews equipment development. North American Controls' objective is to develop and implement process control systems to automate setworks positioning at individual machine centers against optimized sawing and cutting solutions. In theory we can predetermine the optimum log sawing solutions and determine the precise trim length for lumber. In practice the operator is incapable of determining precise trim length at 50 pieces per minute within the grade rule restrictions on wane allowance. Technology is available to do this. The ultimate goal at North American Controls is to develop real time information systems that allow production managers to measure operating efficiency by comparing theoretical log-to-lumber recovery factors with actual results for any log size distribution. The operation of scanners and computers is discussed, with emphasis on determination of optimum sawing sequences and their relation to recovery.

The trimmer performs three basic functions in a sawmill: (1) it squares the ends of the lumber, (2) standardizes length and (3) removes defects to raise grade. Neely cites losses as high as $350 per shift due to trim error only. At another mill, trim error was said to represent 7 percent of the volume of lumber trimmed to remove defect, with an estimated 30 percent of the production being trimmed for defect removal—a loss of a little over 2 percent on total production. Trim errors result from a number of interacting factors, including the combined difficulty to visually quantify defects; the complexity of the decision; the speed at which trimming decisions must be made; and perhaps, to some degree, operator apathy. Trimmer control packages can be developed that make trimming decisions based on quantitative measurements of wane. These systems will provide a means of reducing losses due to trim errors. The lumber tally system that can be incorporated into the trimmer control system, or that can stand alone, will provide necessary information for inventory, production, marketing and sales management.

CHAPTER 16   **CONTINUOUSLY RISING TEMPERATURE DRYING**
By Dr. Dallas S. Dedrick

DR. DALLAS S. DEDRICK, lumber drying consultant to Irvington-Moore, Portland, Oregon, and writer for *Western Timber Industry*'s monthly column on wood drying, graduated from Oklahoma City University in chemistry in 1927. He taught and did research at North Dakota State, Iowa State and the University of Oregon before joining Weyerhaeuser's research division in 1944. Since his retirement from Weyerhaeuser, he has served as consultant to Irvington-Moore on the constantly rising temperature (CRT) drying process, in the invention and development of which he played the key role while with Weyerhaeuser. Dedrick is interested in surface and interfacial tensions in wood and wood/water relations. A member of the American Chemical Society, the Forest Products Research Society, the West Dry Kiln Club, and the American Society for Testing and Materials' wood/water relations committee, he is also a fellow of the Society of Wood Science and Technology. He belongs to several honorary societies. Dedrick is chairman of the Cowlitz County Comprehensive Health Planning Committee and of the regional executive committee and is past chairman of the Cowlitz County Republican central committee.

CHAPTER SUMMARY: Continued study of fundamental physical phenomena—such as freezing, boiling, evaporation, swelling and shrinking—will, in the last analysis, yield the most useful information. Development of the constantly rising temperature (CRT) process for drying lumber is traced—from over a decade of experimentation by a Weyerhaeuser Co. team to patent purchase by Irvington-Moore. CRT is a controlled heat transfer causing a constant rate of moisture removal. CRT (1) takes substantially less kiln time, (2) can be a comparatively low-temperature process, (3) decreases strain on heat sources and (4) cuts falldown in grade.

In conventional drying, moisture loss decreases with time as the wood heats. CRT maintains a constant gradient between air and wood temperatures, making moisture loss constant. Dramatic cuts in kiln residence times with CRT drying are cited. Starting temperatures are in the 90- to 120-degree range and have been as low as 60 or 80 degrees. Decrease in degrade with CRT drying has shown increased dry values of $6 per thousand board feet in Douglas fir and southern pine, when compared with high-temperature drying.

CHAPTER 17   SURFACING DRY LUMBER BY SANDING
By Sherman Kirchmeier

SHERMAN KIRCHMEIER, plant manager of Louisiana-Pacific's Oroville, California division, worked as a log scaler while putting himself through Chico State College, from which he graduated in 1962. He became a scaler for F. M. Crawford Lumber Co. in 1956 and was promoted to forester for the Potter Valley mill early in 1966. He then moved to land manager and plant manager for the subsidiary Apache Lumber Co. mill before becoming plant manager at Aborigine Lumber Co. in 1968. Georgia-Pacific Corp. bought both Crawford and Aborigine that year. In the summer of 1969, Kirchmeier was transferred to G-P's Feather Falls mill on the Sierra side of California to take responsibility for completion of the new mill and to close the old operation there. He was named manager of the Oroville division of Louisiana-Pacific at the beginning of 1973, upon the corporate splitting of G-P's system. Kirchmeier is a director for District VII of Western Wood Products Association and is on WWPA's technical committee. He is president of the Oroville Chamber of Commerce and is vice president of that community's industrial development corporation. He is active in the Boy Scout movement.

CHAPTER SUMMARY: Better wood utilization is necessary to mill survival; abrasive planing can aid utilization. Two types of abrasive finishing systems—batch and in-line—are discussed.

Louisiana-Pacific has a batch system at its Oroville, California plant. It involves a dual line with two-headed edge planes, accumulating lines and a four-head, 52-inch-wide abrasive planer or sander. After edge planing, boards fall onto an accumulator, which lines them for entry into the sander. Capacity is 12 2x4s. If removal keeps pace, the belt can feed 125 feet per minute (1,625 board feet)—an advantageous speed.

Louisiana-Pacific has an in-line system at its Samoa, California plant. It dresses top, bottom and both edges, one board at a time. Profiling knives and saws can be mixed with the abrasive heads. The system can use the space of the present knife planer and its offbearing system.

The belt sander dresses to close tolerance, causes no knot damage or grain tearout. It gives high production and yield with low maintenance.

Abrasive planing requires close sawing tolerances. It is not economical with heavy sawing that has chipped out or torn grain. The Oroville mill targets much closer cuts without grade loss—especially in shop lumber. Crooked lumber is no problem. More durable grits are expected to cut future costs and to permit sanding of green lumber.

CHAPTER 18   SETTING THE MILL TO MINIMIZE MAINTENANCE

By Bob LaBelle

BOB LaBELLE is a sawmill engineer with the U.S. Plywood Division of Champion International, Bonner, Montana. A 1950 Michigan State University graduate, LaBelle gained a wide background in the wood products industry by scaling, saw-bossing, clerking, running lines, laying out roads, cruising, supervising a Prestolog plant and working as plant foreman and mill engineer. In 1964 he joined U.S. Plywood Corp. (now Champion International) as project engineer for lumber at its Redding, California plant. In 1972 he was transferred to the same post at Bonner, Montana, following Champion's purchase of the Montana timberlands and plant of The Anaconda Co. Refurbishing that mill has become LaBelle's responsibility.

CHAPTER SUMMARY: Good maintenance requires familiarity with the work crew as well as with the physical plant. The in-line nature of sawmill production makes breakdowns costly because they shut off production. No maintenance program will totally eliminate downtime, but should minimize it. New equipment will not solve the problem; poorly maintained, it quickly becomes junk. The results of ignoring the necessity for constant and careful maintenance can be costly.

A planned maintenance program should be realistically presented to management in terms of start-up cost increases and long-term cost savings. Supervisors must be convinced it will increase production or make their jobs easier. Include supervisors and millwrights in program planning. Give production-line people some responsibility for machine maintenance decisions. Rotate maintenance people to hold their interest and give them additional skills.

Keep simple records, requiring minimum daily entries. Records should include manufacturer's file, index cards for electric motors and gear boxes, and lost-time records on all machines. Diary-type books can best record and schedule preventative maintenance jobs. Maintenance checklist should be developed for each man/shift. Lubrication records, schedules and checklists based on manufacturers' information are essential; they make the oiler, if properly trained, the best maintenance inspector.

Two qualities essential to a successful maintenance program are versatility and simplicity. Constant analysis and updating will improve the program.

CHAPTER 19   CAN CULLS BE CASHED?

By H. B. "Dutch" Gram & John Gram

H. B. "DUTCH" GRAM, president of Gram Development Co., Portland, Oregon, set chokers at the old Inman-Poulsen lumber mill in Portland to help put himself through the University of Oregon. In 1932 he organized Gram Mfg. Co. to remanufacture spruce into refrigerator framing at plants in Vancouver, Washington, and north Portland. In 1936 he developed the progressive drying concept (known as Blo-R-Dry) for mills that could not afford masonry kilns. Gram purchased mill interests at Kalama, Washington in 1955, quickly becoming involved in salvage from a disease-kill in adjacent woods. Gram Development Co. has grown from the dreams produced by a lifetime in the woods industry. Since the sale of the Gram Lumber and Forest Utilization mill interests at Kalama, the Grams, through Gram Development, have intensified their efforts to identify and develop feasible uses for wood residue. Despite efforts to classify himself as "semi-retired," Gram is in close consultation with the Forest Residue Reduction Committee, with U.S. Forest Service Region VI and the Forest & Range Experiment Station.

JOHN GRAM, a vice president of Gram Development Co., graduated from the University of Oregon in economics in 1952. Born in Portland, he spent 17 years with Jantzen, Inc., a Portland sportswear manufacturer, primarily in managing distribution and inventory control in Portland and Los Angeles and with one year in sales in San Francisco. On his return to the Northwest in 1969, John joined his father, "Dutch," at Gram Development Co. The firm was then in the salvage logging business at the Kalama operation. John Gram initially served as president of the Forest Utilization Co. Since sale of the Kalama mills in August 1972, he has devoted his time to utilization consulting and research. John Gram has also been active in Republican party affairs.

CHAPTER SUMMARY: Gram Development Co.'s breakdown sawing of culls and attempts to find economic uses for substandard logs are described by John Gram. On the average clearcut, 120 tons of residue are left. Reasons for this residue and efforts to utilize it are analyzed. Attempts to cut low-grade sawlogs and utility on a circle headrig with Rosser head barker are described. Careful measurement of log input and recovery showed the effort a "no go" in the 1969-1971 market. To break even, the processor must get 40 percent of the gross in lumber plus 2½ units of chips per lumber thousand. Some changes in mill crew thinking are required for handling substandard logs. Central processing locations for substandard logs and usable residue proposed.

Dutch Gram notes the lack of progress in using substandard logs since the 1964 publication of "Economic Feasibility of Utilization of Logging Residue." Failure of "per-acre" pricing and inadequacy of government cost allowance for yarding unmerchantable material (YUM logging) are noted.

Better lumber recovery means less chips from mill logs. Japanese chip demand may double in four years. While chip demand may rise, it won't bear major price increases. The U.S. Forest Service estimates that 50 processing stations would be needed to clean up woods residue in the Douglas fir region.

Local energy shortages might be alleviated by electric generation using logging residue—but not regional or national problems. Cost of hog fuel transportation is not competitive with coal.

Sawmill automation—the industry's future.

1

Small log handling & sorting systems

James H. Porter, Chief Engineer, CAE Machinery, Ltd. Vancouver, British Columbia

The intent of this chapter is to provide history and capabilities, components and characteristics of automatic high-capacity, small log handling and sorting systems. It should help mill owners, designers and engineers assess the suitability of these systems for their particular needs.

Log handling and sorting has been an essential part of tree processing ever since trees were first used by man for fuel or shelter.

CAE Machinery, Ltd., Vancouver, British Columbia, my employer, has been involved in the manufacture of automatic systems based on Swedish concepts for four years. The old Sumner Iron Works, a predecessor firm, was involved for many years in producing sawmill machinery.

These automatic systems originated in Scandinavia. There, small-diameter logs have been profitably converted into lumber for many years. The situation in Scandinavia is different from ours, but some of their developments, if modified, can be of considerable value to us. One such development is automatic high-capacity handling and sorting systems for small logs.

About 15 to 16 years ago, the Scandinavians, with their through-line sawmills, found it necessary to process logs very rapidly to utilize their machinery to capacity. Automatic sorting systems are installed now in most Scandinavian mills. Logs are sorted for diameter, species and grade. This enables mills to run with one diameter at a time. Logs flow in close succession. There are no changes in settings of heads or saws on these processing machines between logs. Continuous processing is essential for Scandinavian mills to be profitable, as they operate at lower feed speeds than ours. The slower speeds produce sawn lumber sufficiently smooth and accurate to be acceptable in a market for unplaned lumber. A modern Swedish sawmill, located at Rundvik in northern Sweden, is shown in Figure 1.1. Here logs are fed through an automatic deck system into the mill. The logs go through a scaler and are automatically sorted to diameter. Then they are loaded into large automatic storage decks by front end loaders. Usually there are log turners to get the small end of the log into the mill first. Logs pass through a barker. At the Rundvik mill, logs then go through two chipping canters and are formed into approximately square cants. Small logs run in very close succession. The cants move straight ahead and drop into bins much like those underneath a lumber sorter. When larger logs are processed through the two canters, the saw is retracted and the cants pass sideways. The logs go through a frame (gang) saw. The side pieces transfer over, go through a chipping edger, then come out of the mill into an edge sorter. Center pieces go straight ahead and drop into board bays; then, one at a time, these bays or bins are loaded through stacking machines. They proceed through the dry kiln, then through an automatic sorter for sorting and grading. At this point the lumber is ready for wrapping. The Swedish firm Hammars provided the log sorter and the log infeed system for the Rundvik mill.

Value of diameter sorting

In North America, various investigations show the value of diameter sorting with continuous runs of one diameter through single-line sawmills. One of these investigations was made by James Dobie of the Forest Products Laboratory in Vancouver, British Columbia, who reported his findings in the January 1970 issue of *Forest Products Journal.* The findings indicate that "for 1,500 logs of each of three diameters tested, the number of chip-

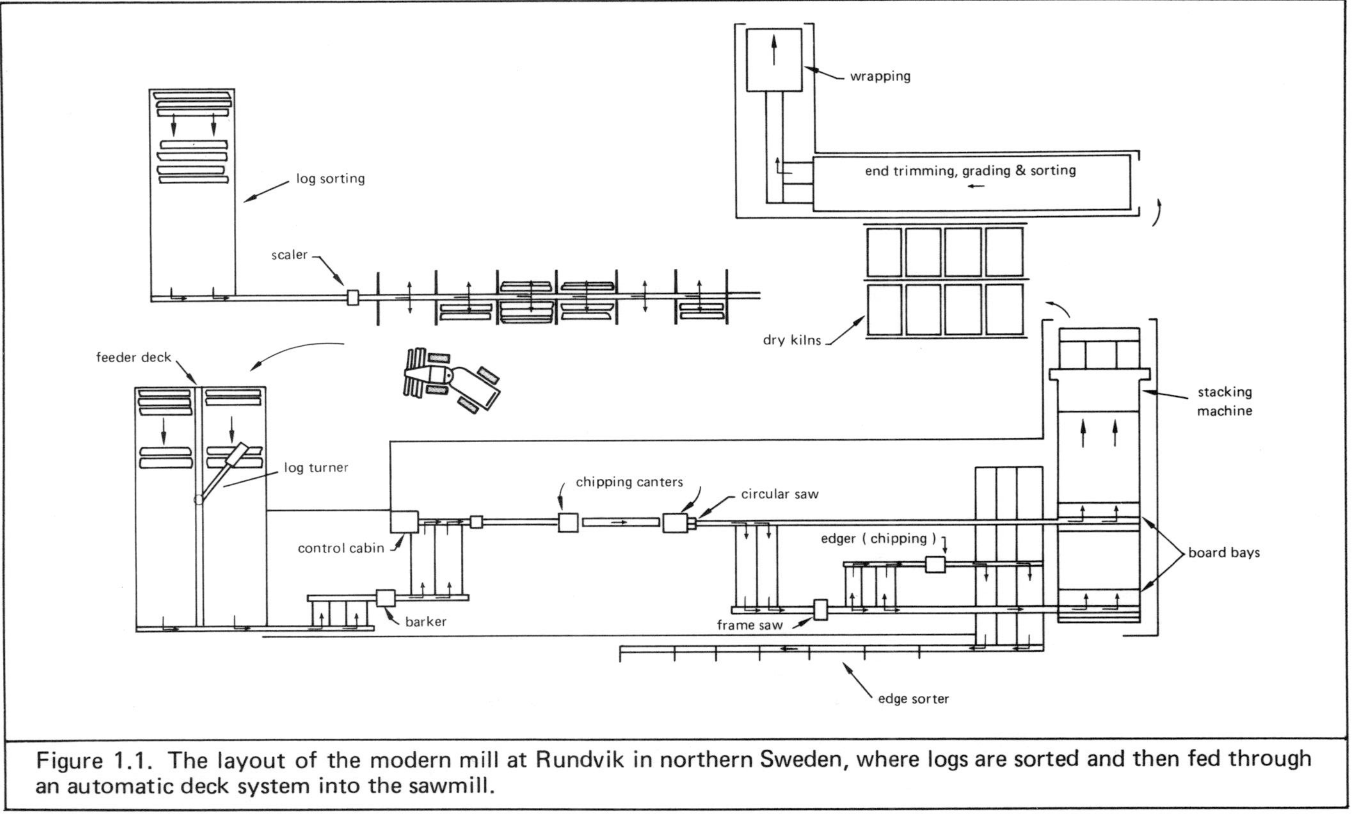

Figure 1.1. The layout of the modern mill at Rundvik in northern Sweden, where logs are sorted and then fed through an automatic deck system into the sawmill.

perhead adjustments required decreased by a minimum of 867 from the no-sort to the sort situation. The increase in production from sorting rose from 11 percent to 153 percent as the average length of gap between logs eliminated by sorting increased from 3 to 30 feet." (That is to say, if the spacing between logs is reduced by as much as 30 feet, it is possible to achieve a 153 percent increase in production.) This is illustrated in Figure 1.2. Dobie's calculation shows that if the gap is reduced from 15 feet to zero, production rises 75 percent. If the gap is reduced by 12 feet between logs, there will be a 60 percent increase in production. These figures are the result of a theoretical analysis and should not be taken as a set condition for any given installation. An analysis should be made upon the logs and the proposed system.

To achieve maximum machine utilization, it is also necessary to design the tail end of the mill so that it is capable of handling a varying piece count. All mill components must work as a single efficient unit if maximum profitability is to be achieved.

Cooperation with Sweden

Over the past 14 years, Hammars has designed and built some 50 automatic log handling systems. CAE Machinery, Ltd., in cooperation with Hammars, has developed nine systems especially to suit North American conditions.

Log merchandisers

Another use for log sorting is with a log merchandiser, a combined bucking and sorting system. In North America, where tree stems are cut to length at the mill, automatic log sorting is being developed as an integral part of the bucking system. Logs are sorted after bucking for optimum use. Separations (other than species) include:

1. Pulp logs (for chipping).
2. Peelers.
3. Chipping headrig or small log mill.
4. Various diameter and length groups for saw logs (large log mill).
5. Particleboard.

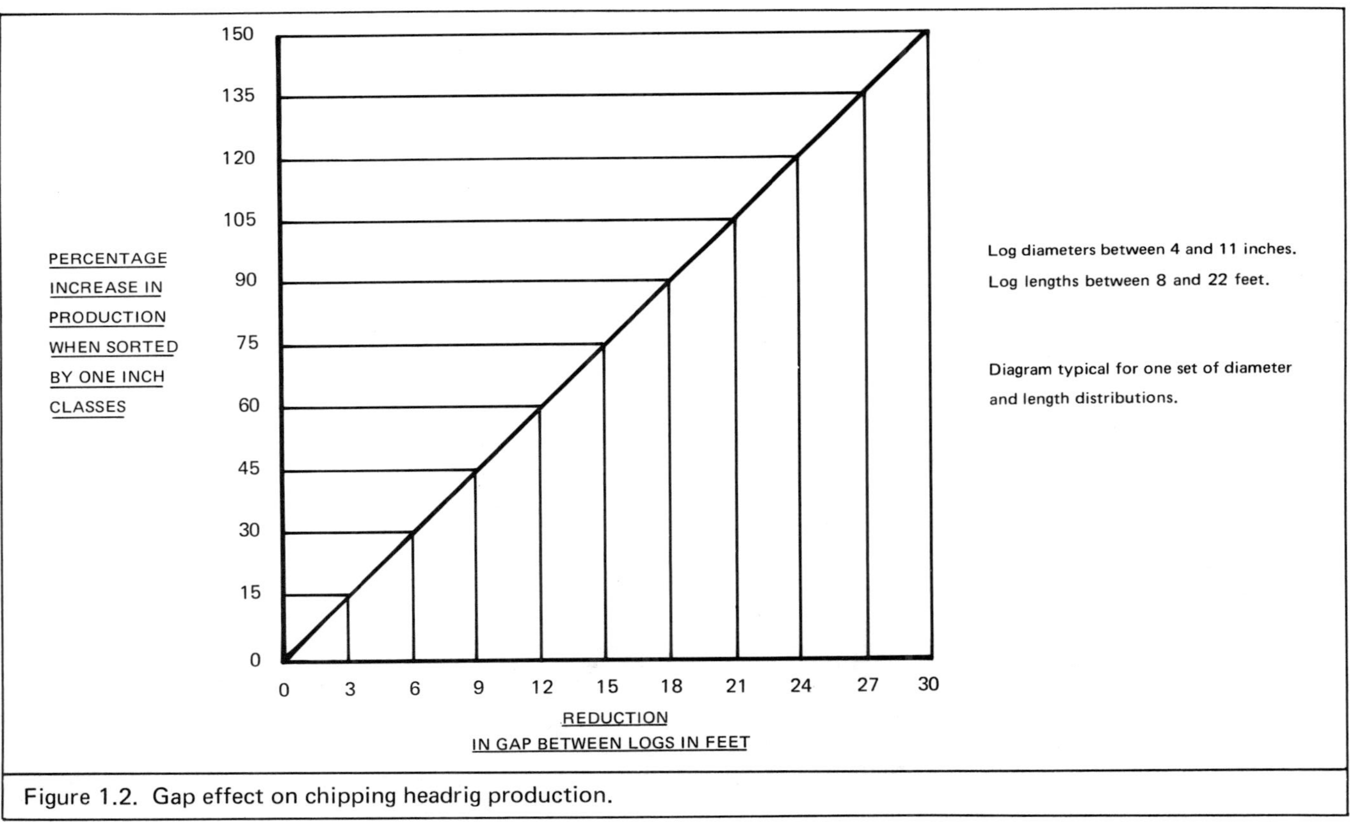

Figure 1.2. Gap effect on chipping headrig production.

Application criteria

A sorting system is only one element in converting logs to lumber. The systems approach must be applied to the whole process to be effective. Log sorters may perform satisfactorily both mechanically and capacitywise; yet, if all other parts of the sawmill are not designed to take advantage of this, the economics of log sorting may not be realized. Techniques such as lineal programming and simulation may be usefully applied to assure better knowledge of the overall system prior to making the investment.

As with most designs related to sawmills, input of the people who will operate the system must be obtained before the designers specify the system to the manufacturer. The operators must be in agreement with the designers that the philosophy of log sorting should be applied in their specific mill.

Often the designers have a good idea, but they fail to get it across to the operators. After all, the operators have to operate the machine. How convinced they are that it is a good idea often determines how well the system works.

Sorting system components

Logs are a unique product of natural growth. No two are alike. They do not all behave in the same way when subjected to mechanical handling.

A well-designed system should provide the high capacity required with the best possible control of all logs at all times. An occasional maverick log can create a jam that will defeat all efforts at high capacity and smooth, continuous flow. Our philosophy is to have two or three stages to gain this good control.

In an automatic log sorting system, there are five basic components (Figure 1.3):

1. Infeed system.
2. Sorting conveyor.
3. Outfeed (discharge) system.
4. Cleanup system.
5. Control system.

All systems must be designed with correct interrelationships.

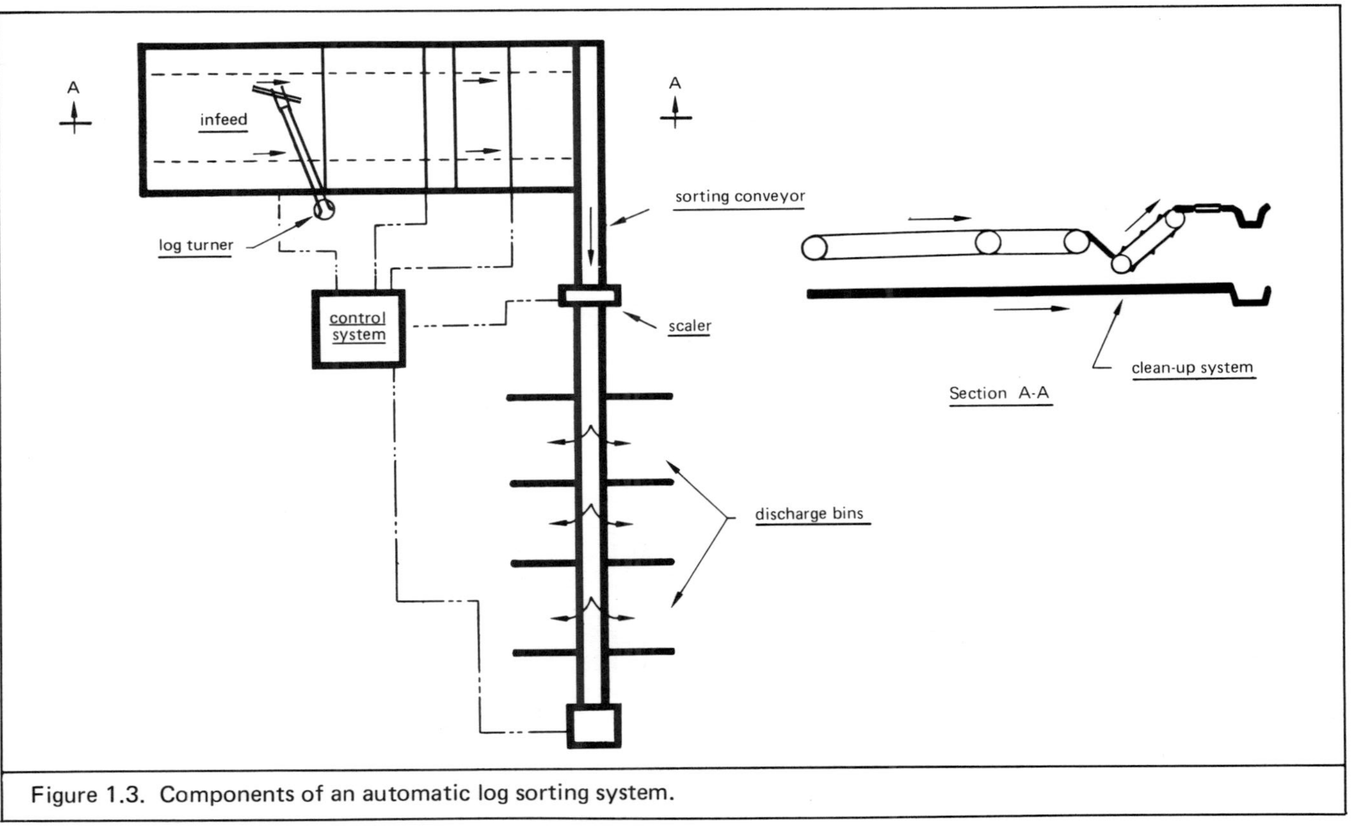

Figure 1.3. Components of an automatic log sorting system.

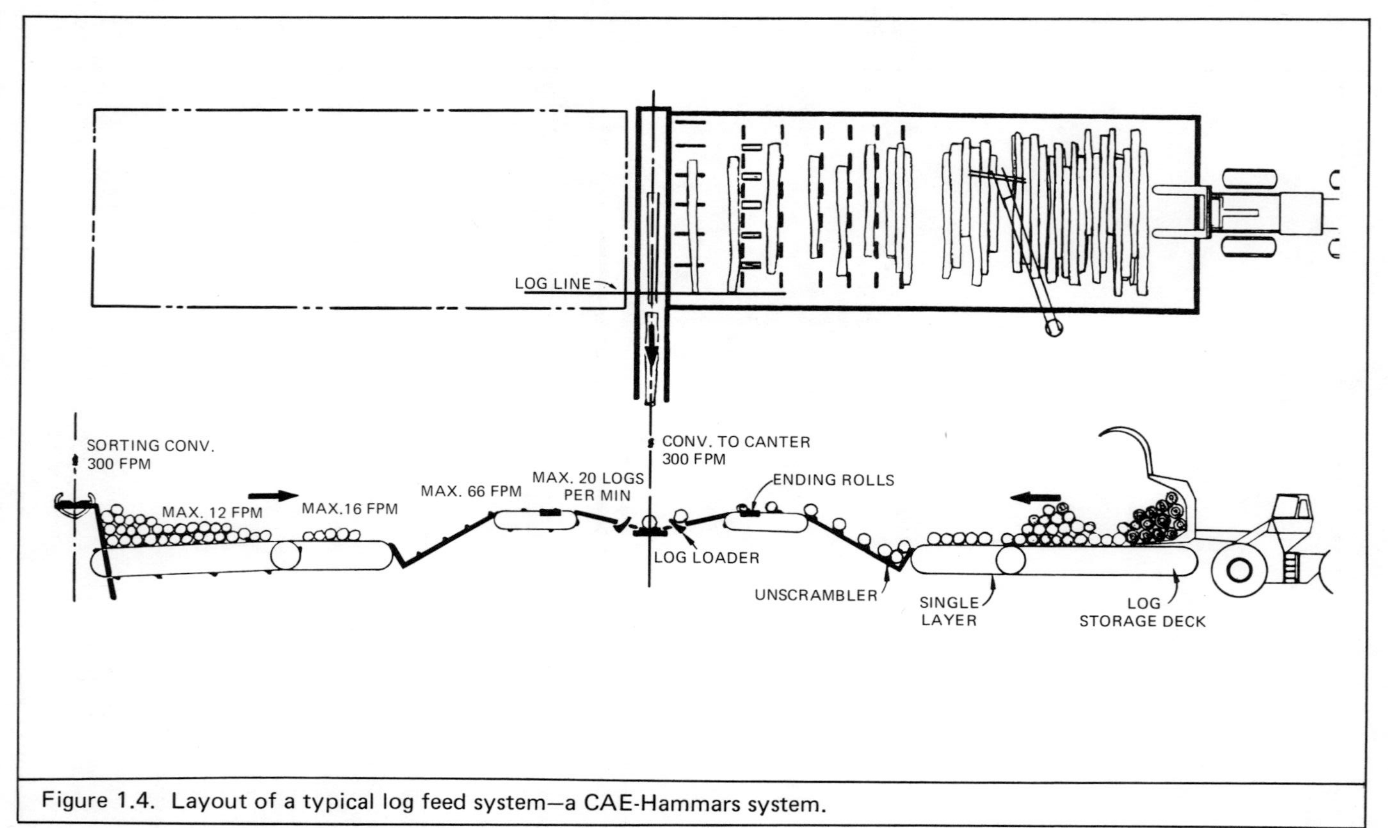

Figure 1.4. Layout of a typical log feed system—a CAE-Hammars system.

Some cleanup system must be provided when sorting the logs before they are barked. Otherwise, bark falls off throughout the system and becomes a significant problem.

Figure 1.4 shows the layout of a typical log feed system (see also Figures 1.5 and 1.9). In this system a knuckle boom turns the logs small-end first. Flow is from the infeed to the sorting conveyor, then to the discharge system or bins. In a typical infeed system (right of center line in Figure 1.4), logs stored several logs high on the storage deck are brought to a single layer automatically to provide a controlled feed into the V-notch of the unscrambler. The unscrambler takes the logs, one at a time, and feeds them to the log loader. Purpose of the log loader is twofold. It provides:

1. A safeguard to guarantee that one log at a time is fed to the sorting system. The unscrambler cannot be guaranteed to provide this safeguard because two logs occasionally come up together. The log loader, however, provides positive separation; it should not be crowded.
2. The means of feeding the logs into the sorting conveyor in close succession. To assist this function, ending rolls bring

Figure 1.5. View of a CAE-Hammars log infeed. The nonslip rubber-mounted log conveyor is at the left.

the ends over to the log line. This gives a common reference line so that, as soon as the tail end of one log has passed that line, the next log can be kicked in behind it automatically.

The left-hand side of Figure 1.4 shows a similar type of arrangement, but here the storage deck is fed from a conveyor and can be automatically loaded. The same feed system applies throughout.

The infeed system was designed to feed sorting conveyors but has proved quite versatile. It is being used to feed sawmills, pulp mills, slasher decks and pulpwood woodrooms as well as sorters. It has good control of the logs, making possible high-speed feeding of logs one at a time.

The single layer deck at Northwood Mills is driven by a hydraulic ratchet drive 1.6).

The small log sorting systems installed for Northwood Mills at Okanagan Falls and at Penticton, British Columbia, are shown in Figures 1.7A and 1.7B. The systems use the basic Hammars principles, as developed for Scandinavia, then modified. A detailed description of the components of the log sorting systems installed

Figure 1.6. A typical hydraulic ratchet drive for log decks.

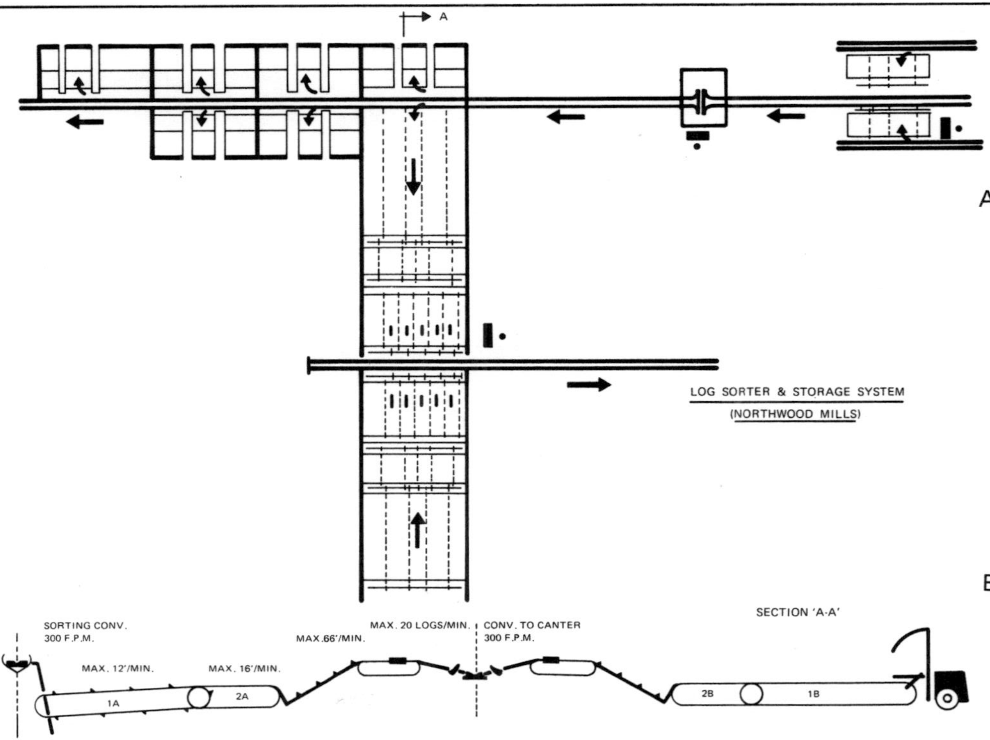

Figure 1.7. Log sorting system installed by CAE and Hammars at Northwood Mills. A. Layout of the sorter and storage system. B. Cross-section of the infeed system.

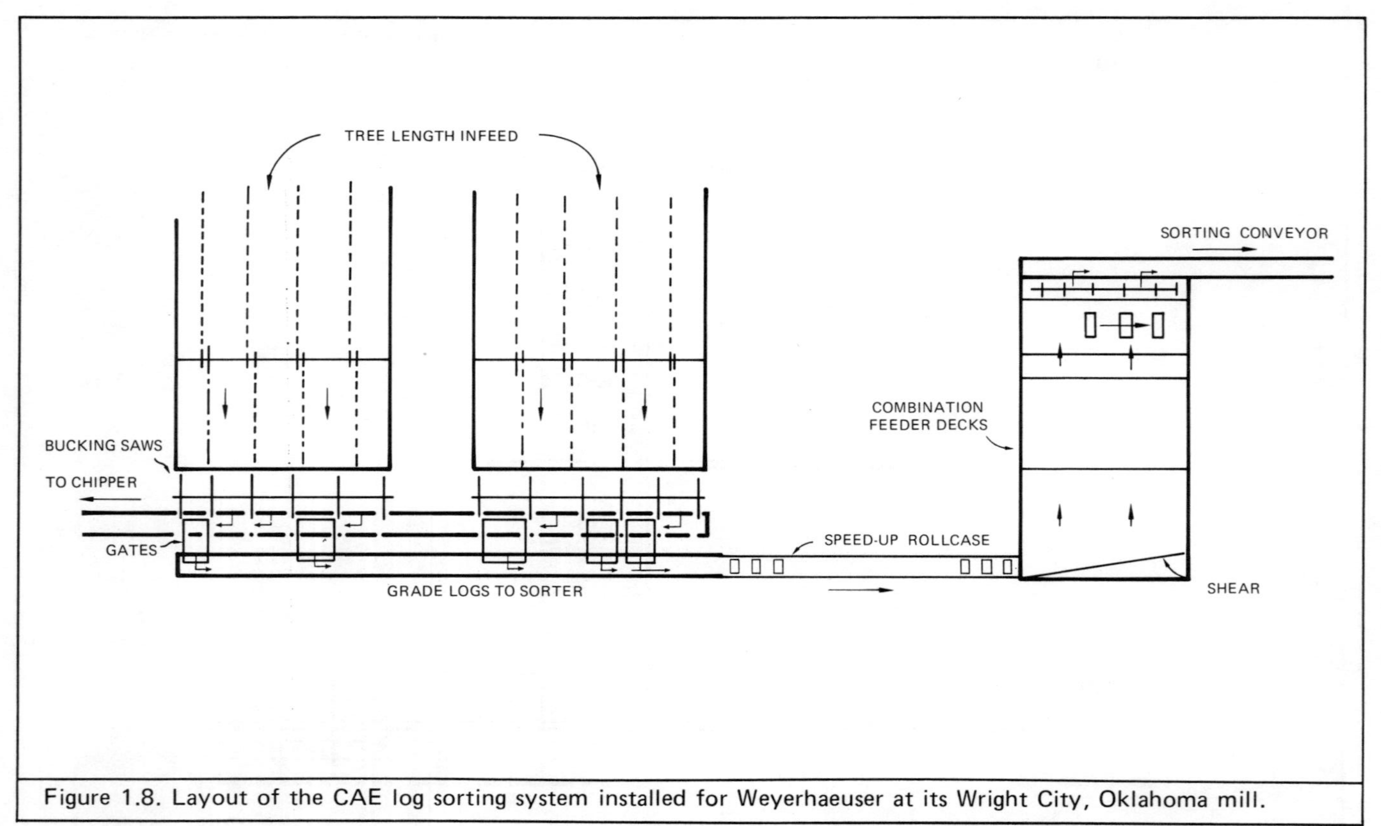

Figure 1.8. Layout of the CAE log sorting system installed for Weyerhaeuser at its Wright City, Oklahoma mill.

for Northwood Mills appears as an appendix immediately following the discussion period at the end of this chapter.

Figure 1.8 shows the Weyerhaeuser Co. installation at Wright City, Oklahoma. The unscrambler is mounted on rubber pads to reduce shock from logs falling into the V-notch, thus increasing equipment life. It is driven by a slow-speed Hagglund hydraulic motor. At the storage and single layer deck, a hydraulic ratchet drive is used. This drive system is good for very slow speed, intermittent feeding; stop-go action is good in stabilizing the big load. It is a very simple drive; there are no reducers—just a straight drive on the end of the head shaft. The cylinders are extended here.

Turning system studies

We have made some studies on turning systems. For very high capacity systems, where 20 logs a minute are required, the best method is to have a knuckle boom remotely controlled to turn the logs on the deck before they enter the mill. There are some mechanical systems that automatically turn logs, but their capacity

Figure 1.9. View of a knuckle boom with log turner at the infeed. The knuckle boom loader, with remote control, is used to turn logs on the deck before they enter the mill. Log are turned so they enter the mill top first.

seems limited to about 12 or 14 logs per minute. A knuckle boom with log turner at the infeed is shown in Figure 1.9.

Merging infeed

On some systems where there is a cut-off saw or a ring barker, the throughput speed of logs is restricted to about 150 feet per minute. A normal high-speed sorting conveyor can take two lines coming from cut-off saws or barkers. Figure 1.10 illustrates a two-line merging infeed. Logs are fed in, transferred together, then fed on a first-come, first-served basis into the sorting conveyor. The kickers transfer the logs off the far conveyor, across and down into the sorting conveyor.

Figure 1.11 is a view taken in a mill at Karlsborg, Sweden; barked logs from two ring barkers merge into the central conveyor, which feeds out to the log sorter.

The sorting conveyor

Next comes the sorting conveyor. The Hammars system uses nonslip flights closely spaced to take any length log. There is no relationship between spacing of the flights and the logs being fed onto them. They are spaced closely enough to stabilize the logs. The sorting conveyor has V-shaped flights with knife edges to transport logs without slippage. Flights run on wear strips and carry the suspended chain, thus eliminating chain wear through friction. Large sprockets and an H-type mill chain give good high-speed performance (up to 360 feet per minute).

Each log is supported by several flights. There is no fixed relationship between flights and logs, as in some other systems. The advantage to this system is that logs can be closely spaced, regardless of length. With shorter logs, more can be carried in the same amount of time.

A shaft mounted below the chainway carries two or three W-shaped kickers; shaft rotation causes the kickers to discharge passing logs. The discharge side depends on direction of rotation.

Kickers operate by a rocking motion to discharge the logs to either side. This provides a good directional kick to the log. It does not kick it up into the air but rolls it off the conveyor, so that it is discharged with good control. Inertia is very low so that the high-

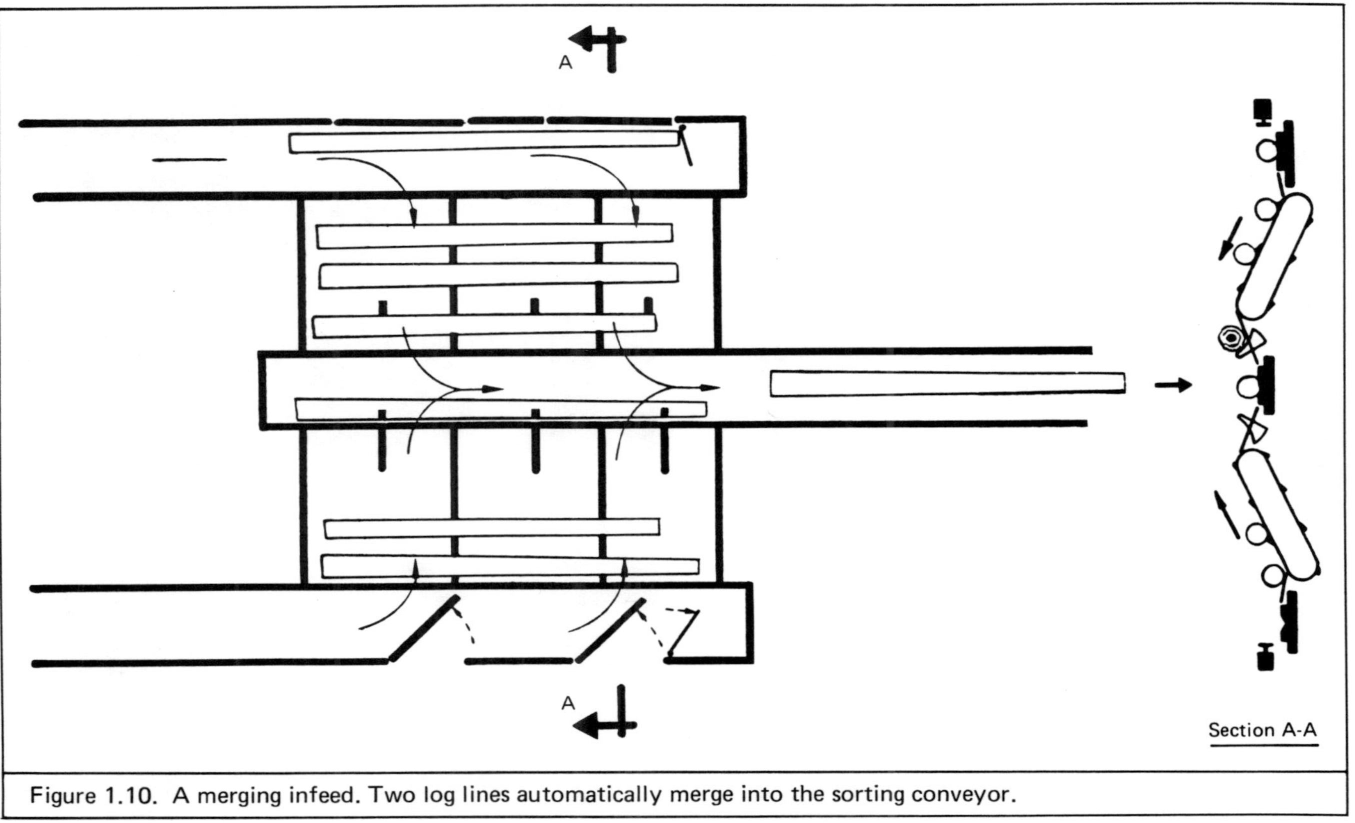

Figure 1.10. A merging infeed. Two log lines automatically merge into the sorting conveyor.

speed kick can accurately discharge the logs into the bins.

Logs may be moving at about 360 feet per minute. After they are kicked off, logs drop onto knife edges that virtually stop forward travel before dropping the logs into bins. Kickers are actuated by a reversing electric motor. Hammars' reasoning in choosing an electric motor is this: An air system is prone to temperature variations. Air being elastic, it is hard to guarantee an accurately timed kick. A hydraulic system is expensive. The cost of an electric system lies between the cost of air and hydraulic, but it provides a very accurate kick. An electric system is consistent, regardless of weather or temperature.

A limitation of the electric kicker is its ability to kick a high number of repetitive cycles into the same bin. It handles about 10 kicks per minute at 65 degrees. A higher rate kicked continuously into the same bin may cause overheating. Studies are underway to develop an electric kicker with much higher capability.

Figure 1.12 shows the infeed to the first sorting conveyor we

Figure 1.11. Barked logs merge into the central conveyor at a sawmill in Karlsborg, Sweden.

Figure 1.12. CAE's first sorting conveyor infeed at Northwood Mills.

made. It is at Northwood Mills. Unfortunately, after a while, the bins were knocked down by the mobile equipment.

Sorter bins

The Swedes have developed very good bins. They married the mobile equipment to the bins so they can unload without damaging the bins, the sorting conveyor or the mobile handling equipment. In the system to be installed for Great Lakes Paper, concrete bins will be used. The Swedish idea is for the log to drop, then lose its energy as it rolls around. The log goes down, across and up the other side of the bin a bit. As soon as there is a pile across the bottom, there is no problem as the energy is absorbed by the lower pile. Shown in Figures 1.13 and 1.14 is a Swedish system with strong, heavy steel bins. Mobile equipment fits closely into these bins to lift the logs. In this arrangement, forks can go under the bin and the grapple can go around and through to get around a log pile. By applying pressure to the forks and grapple, the machine operator can then squeeze the pile and lift vertically, reducing pressure on the bin.

Figure 1.13. Heavy steel bins are used in this Swedish sorting system. The system utilizes nonslip flights and electric kickers.

Figure 1.14. A close-up view of sorting bins.

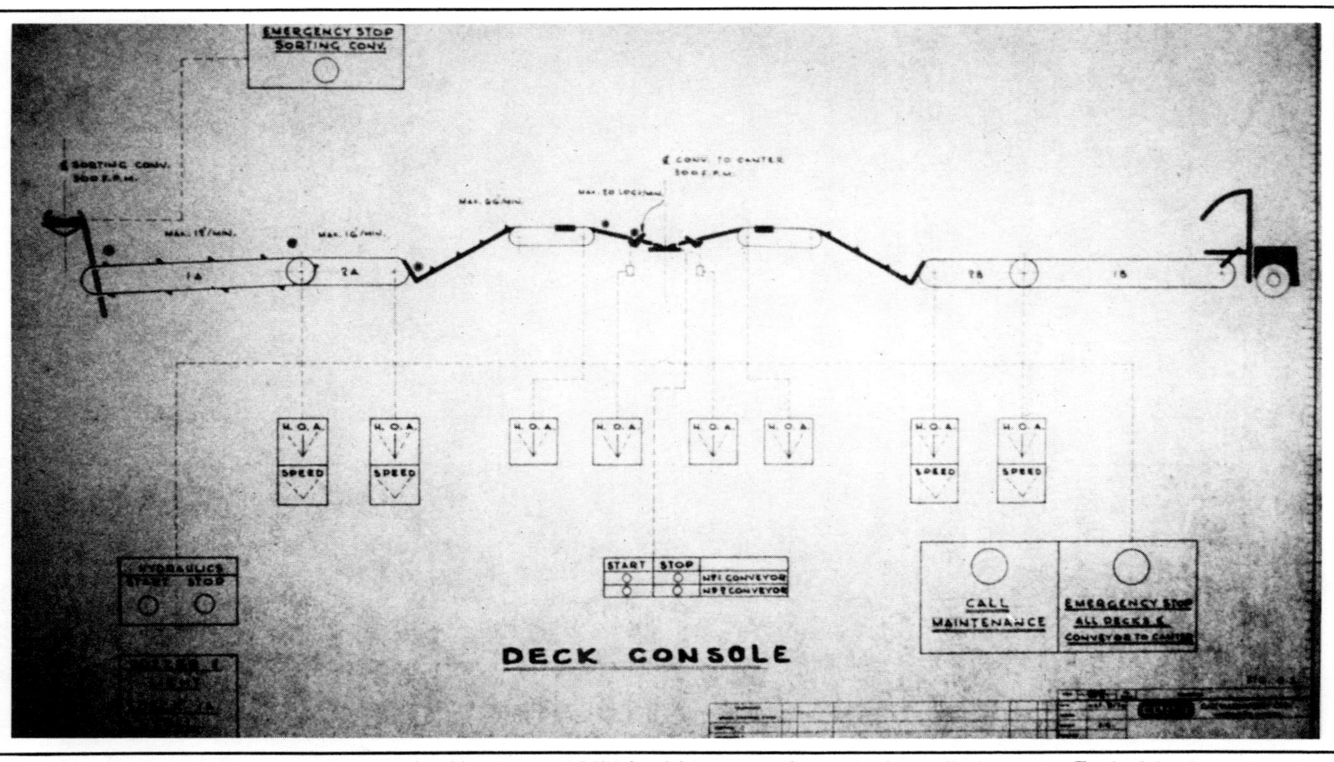

Figure 1.15. Infeed control system (at Northwood Mills) with automatic control on all elements. Each drive has a hand automatic off-switch.

Infeed control system

Figure 1.15 illustrates the controls for a typical infeed system. There is automatic control on all drive elements, but each has a hand automatic off-switch. It appears excessive, but this really is not so. Assuming the system to be loaded, as the log is transported off in the sorting conveyor, it passes a photocell. When that is clear, the log behind kicks in. As soon as that log loader is empty, the log behind moves into the loader. It is a sequence operation down to the V-notch in the unscrambler. In the unscrambler, there is a detector to control the height of the logs. Too much height in the unscrambler may carry too many logs over the top and control is lost. Height is controlled by controlling the single layer deck feed.

In feeding from a sorting conveyor to a loader deck, loading and buildup of the storage deck is also controlled by the building of a pile of logs at that point. When the pile is above a set height, it will automatically move forward a preset distance. Logs load behind that pile, so that the pile builds progressively all the way across the

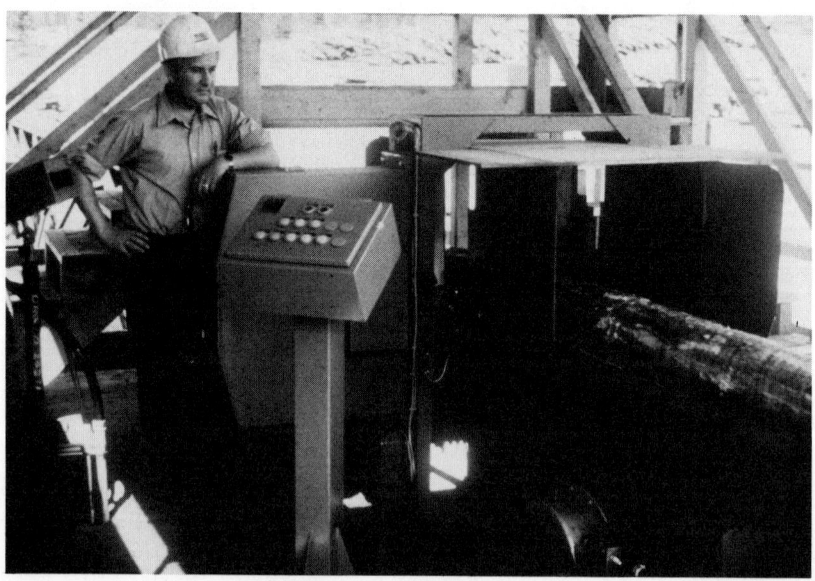

Figure 1.16. The electronic scaler was developed to measure log diameter accurately. The REMA system is shown here.

Figure 1.17. The ASI system for measuring log diameters accurately.

deck. When the deck is full, a warning device notifies the operator, who then must start emptying the deck or stop the infeed.

Sorting conveyor control

CAE and Hammars have found the electronic memory system with a pulse count to be the most successful sorting conveyor control system. Pulse count, controlled by the tail sprocket on the sorting conveyor, is proportional to the distance traveled by the log. Assume a sorting conveyor with 1,000 pulses for a full-length travel. The memory system understands that a log of a specific diameter must discharge half way. It counts. When it has counted 500 pulses, the log is kicked off. The pulse system, which is adjustable, must be short—about two inches.

The log center is determined so that logs are always discharged into the centers of bins to insure stable loads.

Automatic log measuring

How is log diameter measured? In handling 20 logs per minute, it is virtually impossible to judge size accurately. Logs are not smooth enough for simple measuring gadgets. So the electronic scaler (or log-diameter measurer) was developed. Shown in Figure 1.16 is the REMA system, installed at Northwood Mills. Figure 1.17 shows the ASI system. In neither does any mechanical component touch the logs. The Hammars system (using REMA) is limited to measuring diameters to 30 inches at the very maximum.

The limitations

To have good control, the log diameters must not exceed a 5 to 1 ratio. If the maximum is 30 inches, 6 inches is the minimum for good control. But smaller sizes—for instance, 4 to 20 inches—can be handled quite well. Less than 4 inches causes trouble. Log length can be anything from 8 to 24 feet. Capacity varies with log

Figure 1.18. Speed-up rollcase showing portions of log shear, feeder decks and unscrambler at Weyerhaeuser's Wright City mill. Feed speeds of 600 feet per minute are obtained.

lengths. Typical capacity: 780 24-foot logs or 1,500 8-foot logs per hour.

The Hammars system can:

1. Sort with minimum gap processing. More than 9,000 logs have been sorted in an 8-hour shift.
2. Provide or facilitate the mill's maximum processing machinery utilization.
3. Maximize lumber recovery by reducing human error.
4. Improve grade recovery.
5. Cut manpower needs.
6. Reduce wear on setting mechanisms of log breakdown equipment with consequent reduction of maintenance.

Wright City system

A recent installation is a hardwood merchandising system for Weyerhaeuser Co.'s Wright City, Oklahoma mill. Tree lengths are brought into decks. They come through cut-off saws and are fed into the log sorting system with chunks going to the chipper. Bush Mfg. Co. provided the cut-off system; CAE provided the sorting system; REMA furnished the electronic control system for CAE's sorting machinery.

Logs to be sorted arrive in a random way. The speed-up rollcase shown in Figure 1.18 hurries logs to speeds of 600 feet per minute. Logs may come side by side or two or three on top of one another. Speeding helps draw them apart. The logs are dropped on a shear section and transferred sideways. The logs then drop onto a standard combination infeed system, feed out through the unscramblers and into the sorting conveyor and are then discharged farther along.

At Wright City there is also a hickory feedback system. Logs drop onto a deck, transfer and come back through a ring barker. In the control house, the operator makes grade decisions only. There is no diameter sort. As the logs pass by, the operator punches a button, determining their destination.

Sorter system design

These are the basic considerations in designing a sorter system:

1. Insure that all conveyors, loaders, kickers and detectors are suitable for all log diameters and lengths to be handled.
2. Design with flexibility. Separate and feed logs one at a time wherever possible to keep control.
3. Minimize the space between logs—reducing forward speed required to meet capacity objectives accordingly.
4. Allow for an increase in logs fed per minute when shorter logs are run.
5. Discharge from the sorting conveyor so that logs land in bins or decks, insuring stable, balanced loads.
6. Use a positive-type conveyor for the log sorting line with a pulse count memory system.
7. Mount all equipment subject to impact on flexible mounts, where possible.
8. Employ variable speed drives where logs are fed transversely to allow flexibility needed for handling varying diameters to match capacity.
9. Insure adequate storage deck capacity to handle infeed surges.
10. Employ adequate control devices with manual overrides.
11. Use accurate automatic log measuring devices with tested, proven control systems.
12. Employ polarized light for photo detectors so they are un-affected by the sun. These detectors should be mounted independent of any vibrating structure.

Other types of mechanical sorting systems

Gravity discharge type. Two kinds of gravity discharge sorting systems are fairly common. The two types are:

1. The track on one side of the conveyor is lowered to allow the logs to discharge (Figure 1.19, Type A).
2. A supporting arm attached to the log-carrying chair or flight retracts to allow a log to discharge (Figure 1.19, Type B).

In both these cases, each log is carried by two or more chairs or flights with gaps between logs varying as the spacing between log centers is based on the longest log. Carrying chairs are located and

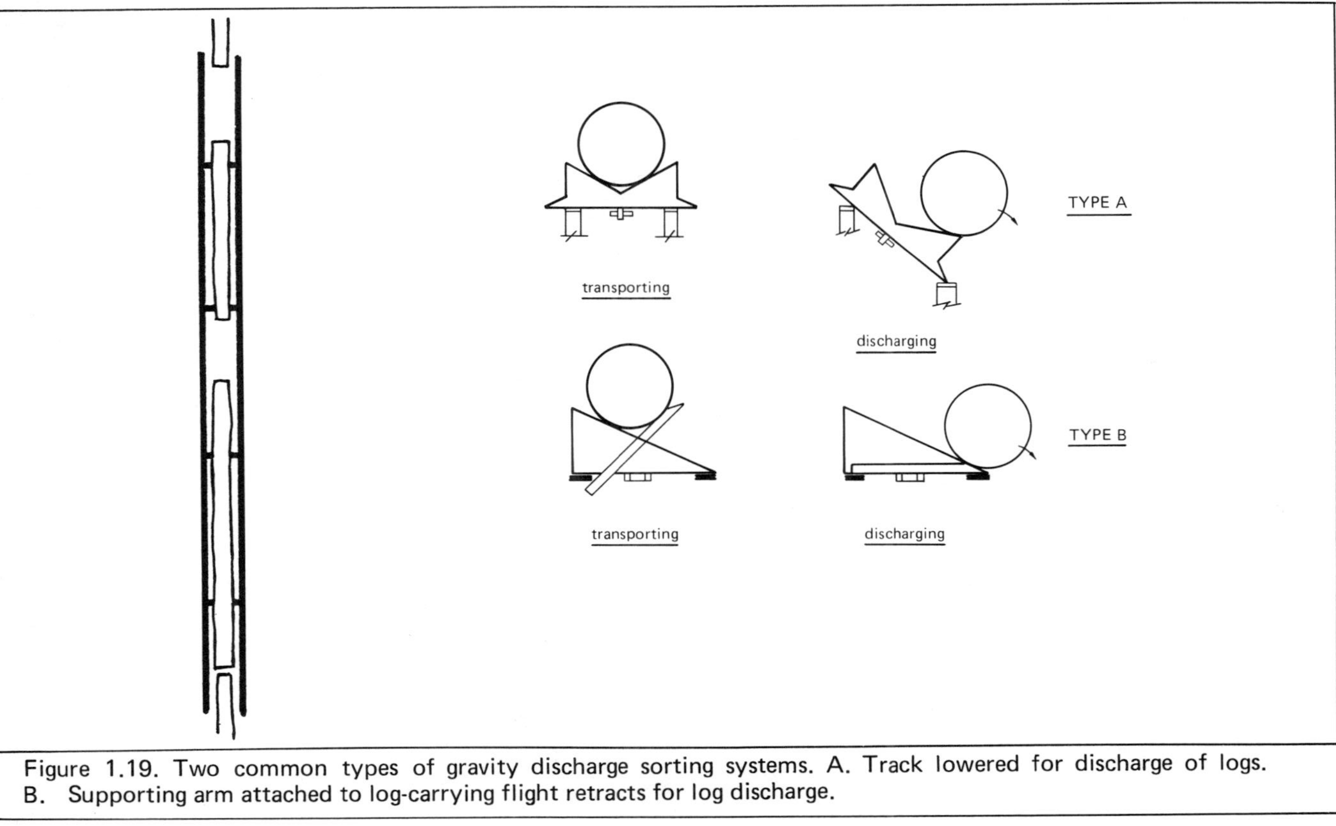

Figure 1.19. Two common types of gravity discharge sorting systems. A. Track lowered for discharge of logs. B. Supporting arm attached to log-carrying flight retracts for log discharge.

spaced on this basis. Capacity is thus limited.

Concave rollcase sorting conveyor. Over many years, concave rollcases have proved a good log-conveying method. Applied to log sorting, however, they are at a disadvantage, as logs of varying diameters move over the rollcases at different speeds. So it is necessary to use limit switches or other detecting devices along the length of the sorting conveyor to control the discharge. If kicking is to both sides, detecting devices must be located between the rolls, beneath the logs. Detectors are subject to repeated malfunction. This is due partly to uneven log surface, which results in unpredictable operation. It can also be caused by wood rot and bark getting into the mechanism.

Chain and belt conveyors. Sorting can be performed by chain and belt conveyors. If logs ride directly atop chains, the initial cost is lower than for a rollcase. However, conveying is not positive enough to eliminate the use of locating detectors along the conveyor length.

Transverse and L-bar sorters

So far our attention has centered on systems in which logs move in line, one behind the other. Such systems are by far the most common for sorting logs. However, consideration should be given to possibilities of sorting where logs move transversely. Two such systems are described below.

Logs move transversely across deck; gates discharge to lower level rollcases. This system is good for logs that move slowly and are tree length or up to about 60 feet long. A disadvantage is that it is very difficult to develop gates that will operate satisfactorily at high speed, both to carry logs across and to discharge the logs as needed.

L-bar sorters. These are used in fairly large numbers for sorting lumber. L-bar sorters can be modified to handle small logs. Such a system has high production capability. It can possibly handle in the order of 45 logs per minute. This is more than double the capacity of an in-line sorting conveyor. However, the L-bar system is expensive because all equipment must be suspended well above the ground. Rugged bins, carts or conveyors must be provided to accept logs as discharged.

DISCUSSION PERIOD

MILTON MATER, president, Mater Engineering Co., Corvallis, Oregon: We have found that using the pulse systems connected to the sprocket for log kick-off resulted in a good deal of error because of logs bouncing on the chain—particularly misshapen logs. Are you counteracting that?

PORTER: First, we allow the log to settle down. We have a short length of conveyor that the log drops onto where it gains speed up to the velocity of the sorting conveyor. The conveyor is designed so that flights move along very smoothly. It is not a conventional mill-type chain. We have V-shaped flights on wear strips on either side with chain suspended beneath, so the action is very smooth and the log is traveling almost as though on a railway car. There is no vibration because the track is smooth and level. Certainly, we have had no problem with slip and neither has Hammars with its 50-or-so installations.

DENNIS OLDKNOW, Phillips, Barratt, Hillier, Jones & Partners, Ltd., consulting engineers, Vancouver, British Columbia, addressing Joseph Dixon, construction engineering manager, Weyerhaeuser Co., Tacoma, Washington, a user of the CAE-Hammars system: What is your opinion of the practical operating characteristics of the Wright City, Oklahoma installation. Particularly, how many logs are you actually getting through the Bush system and the Hammars sorting system?

DIXON: We have three Bush systems. Two are operating on pine and one on hardwood. Our hardwood merchandiser is primarily oriented toward providing a third of the chip supply for an Oklahoma pulp mill. We have no other current outlets for our hardwood grade segments except for log sales. We have run tests on the system, using pine, that met the design criteria, which was 20 stems per minute. This is a 6-sort pocket. We have never put it to the test in hardwood, except to fill the surge pocket and then run it. The system has met the 20 stems per minute (or 20 sorts per minute) coming out of there. We are taking out hickory, which is very hard to bark in a drum debarker, with the grade, even the hickory fiber. We then use one of those sort pockets to kick the hickory onto, then take it through the ring debarker and a hardwood barker. We intend to average 10 stems per minute when this thing is really on the line. I think it will do that. It will hit peaks of 20.

GUS HALEY, general manager, M&R Timber Co., Port Angeles, Washington, addressing Jim Porter: How many sorts were you making? And, in round dollar figures, what is the cost of the system?

PORTER: At Wright City, we had 12 sorts, 6 on either side. It is difficult to put that into figures because our electric is included. Figures will also vary with the amount of installation included. There were many take-outs. Weyerhaeuser itself provided many little pieces—chains, hydraulic powerpack, hydraulic cylinders. We provided the control system, but they furnished the motor control center. For the part we supplied, today, very roughly, the figure is $200,000.

DIXON: I would agree with that figure, very roughly.

APPENDIX TO CHAPTER 1

A description of small log sorting systems, installed by CAE and Hammars, follows. These are for Northwood Mills at Okanagan Falls and Penticton, British Columbia.

Log sorting systems at Northwood Mills

The log sorting systems installed for Northwood Mills, Ltd. at Okanagan Falls and at Penticton, British Columbia, use the basic Hammars principles, as developed for Scandinavia, then modified. The requirements were set by Brodie Gillies, Northwood's chief engineer, with Phillips, Barratt, Hillier, Jones & Partners, Vancouver, British Columbia, consulting engineers.

Requirement is to sort logs by small-end diameter into six sizes: 4 to 4 7/8 inches; 5 to 5 7/8; 6 to 6 7/8; 7 to 7 7/8; 8 to 11 7/8; and 12 to 16 inches. Lengths vary from 8 to 24 feet. Capacity required was infeed up to 15 logs per minute with outfeed up to 20 logs per minute.

Infeed: Logs, barked and sorted to length, flow in two lines into the system. Lines are automatically merged into one sorting line on an intermittent feed basis. Automatic infeed has manual override controls.

Sorting conveyor: Speed of the sorting conveyor is 300 feet per minute.

Scaling: REMA scaling system is used, initially arranged for automatic top diameter measuring only.

Discharge: "Hot" sort is discharged onto a transfer deck where log storage is automatically built up. Thence, automatic feed onto the discharge conveyor is fast. Logs that are discharged into the storage bin can be transferred by mobile equipment into a second storage bin and feed deck. Thence they can be fed automatically into the discharge conveyor.

Discharge conveyor: The discharge conveyor feeds logs at 300 feet per minute into a canter. Spacing between logs is close.

Operators: One operator overlooks sorting and storage process, policing and controlling log feed from storage decks into the discharge conveyor. Merging infeed is policed by cut-off saw operator who, in an emergency, can operate override controls.

Figures 1.7A and 1.7B show the sorter and storage system flow at the Northwood Mills installations.

2

Sateko log sorting system

Robert E. Vadnais, President, Totem Equipment Co.
Seattle, Washington

Dick Johnson, Construction/Maintenance Superintendent
Van-Evan Co., Missoula, Montana

The smallest log stacker Totem Equipment Company sells is its 80,000-pound unit and its stackers range up to 120,000 pounds. Before sorters were developed, mill men simply picked up incoming loads of logs with a stacker and set them onto the ground, into the cold deck or on the mill infeed. Then mill operators started to sort logs, often using a huge $120,000 log stacker—an extremely expensive process.

So about seven years ago, Totem began experimenting with infeed deck design. We not only designed infeed decks, we also put the logs on a moving deck with a chain saw hanging down. The operator had to turn around and try to cut the log, then drop it onto a sorting track. With all this exercise, cost proved prohibitive. In fact, cost was staggering.

I do not remember how I began to correspond with the president of Sateko—perhaps through a magazine article. However, I went to Europe in January 1969 to look at Scandinavian sorting systems. The reason for the January trip was that all of us had bought equipment in June and tried to run it in December with little or no success. Once previously I had gone to Sweden and bought a tractor—only to find, when home, that in Sweden they log laterally along hills, whereas we log up and down. I could get the tractor down a hill, but could never get it back up. So the tractor is in my barnyard now—a very expensive mistake.

Dean Prater, who was then Northwest logging manager for Crown Zellerbach Corp., went with me. The two of us spent about three weeks in Sweden and Finland looking at sorting systems. Without question, we could adapt the Sateko system to the Ameri-

Figure 2.1. The model Sateko log sorting system, designed to handle all logs horizontally until they are placed on sorting chairs.

can market. So what we will discuss here is the Americanization of the Sateko sorting system. It has been painful; it has taken time. But with the help of the Evans Products Co. people and, particularly, that of Dick Johnson, we built the first Sateko sorting system at Missoula, Montana.

Sorting model erected

The model Sateko log sorting system shown in Figure 2.1 is not an exact duplicate of the system we built for Evans Products Co.'s Van-Evans mill at Missoula. The model resulted from our effort to handle all logs horizontally until they are placed on sorting chairs.

The system in Montana has a three-section infeed deck, a cross-feed conveyor and a surge deck. An overall view is shown in Figure 2.2. The surge deck was needed because logs had to be endlined so that we could center them onto the passing chairs. The only possible way to endline those logs without using power rolls was to put them onto a crossfeed conveyor and butt them against a steel plate, creating a fixed zero line from which to measure. Then the centering device with a memory system would center logs on the passing chairs and head them down the track. But this entailed an expense of approximately $60,000.

Figure 2.2. An overall view of the Sateko log sorting system built for Evans Products Co.'s Van-Evans mill at Missoula, Montana.

Figure 2.3. The 20-foot surge deck at Missoula with the crossfeed conveyor at left. The Evans system runs at 300 feet per minute and is capable of sorting seven logs per minute.

To overcome this, we approached Boeing Company at Seattle for someone versed in electronics. If you find the right man, it is easy to solve a problem. I explained that we wanted to find the center of a log. The man from Boeing said that it was really quite simple. Start one set of Logic at a speed of $X + 2$; start another set at $X + 1$. When the original set is at the end, the second set is halfway through and there is the center of the log. Now tie this information into the passing pair of chairs. It is that simple, the Boeing man said. It sounds easy, but how were we to go about doing it? The Boeing representative not only did it, but he built the model in three weeks.

Circuitry solid state

Inside a gray control box in the model is the solid-state circuitry. There are two basic types of solid-state circuitry: commercial and military. Commercial circuitry is good to temperatures of about plus 20 degrees Fahrenheit; military circuitry is good to about minus 50 degrees Centigrade. We have used the military system, and reliability is quite good. When you trigger the mechanism to find the center of the log, you also measure the length of the log, which is a bonus.

The model was built to test the electronic centering device so that we would be able to eliminate the crossfeed conveyor and the surge deck, which eliminated approximately $60,000 of cost, and one man. At some future time, it will also allow us to saw logs on the infeed deck and still be able, with the three-section feeder, to center those logs on the passing set of chairs without endlining them.

Evans system installed

The Evans system at Missoula consists of a three-section infeed deck, oversize log kickout, crossfeed conveyor, 20-foot surge deck (Figure 2.3), centering device and the sorting track itself with 22 40-foot pockets. The system presently runs at 300 feet per minute and is capable of sorting seven logs per minute provided there is sufficient material available to run the line at this speed.

Figure 2.4. View of the "picket fence" log reject.

The problems in putting the system in operation were not minor. We are taking camp run material. Some trees are not delimbed; there are crooked and broken logs. Notice in Figure 2.4 the "picket fence," which is the log reject. When that gate slams down, we can pick a log out of the system. In Figure 2.3, note the centering device that centers a log onto the passing pair of chairs.

Figure 2.5 is an end view, looking at the waterfall, while the system was being built. Figure 2.6 is a view from the control house, looking at the waterfalls as they come across. The combined length of the decks is 87

Figure 2.5. An end view of the Missoula mill, looking at the waterfall. The picture was taken during construction of the Evans system.

Figure 2.6. View from the control house at the Van-Evans mill, looking at the waterfalls as they come across.

feet and the decks are 32 feet in width.

Here is where the problems started. Highway truckloads average 60 logs. Often 8-foot and shorter broken pieces are included. The difficult job is to try to start unscrambling the logs as they cross the deck. When two pieces go over the waterfall at once and other logs get on top of them, the whole group starts to jackstraw, as shown in Figure 2.7. A large short chunk can start a problem.

Log action irregular

The other problem is this: Log stacker forks are 8 to a maximum of 10 feet apart. When a stacker picks logs out of a cold deck and starts across the mill yard, the logs immediately want to jackstraw. When an operator starts to lift the logs onto the deck, action of the outside logs makes him want the infeed deck to be 80 or 90 feet wide instead of the conventional 32 feet allowed here. This cannot be overcome by trying to use force. Logs will come in poorly limbed, larger than the system can take and with more sweep or crook than the system can handle. Since the decision has been made not to try to buck as part of this program, the crook cannot be handled by sawing. So the options are to reject these logs over the end of the infeed deck at the log rejector or, as

Figure 2.7. Failure to unscramble logs as they cross the deck results in a jackstraw. A large short chunk can start a problem.

Figure 2.8. A hydraulic crane picks out the logs that are too difficult for the sorting system to handle.

shown in Figure 2.8, to put in a hydraulic crane to pick out the logs that are too difficult for the system to handle.

We believe that all the controls can be put inside the hydraulic crane so the operator can also do whatever log straightening is needed from there. The crane is certainly capable of putting a lot of logs back into line, taking logs out of the system that cannot be handled and straightening the deck.

Figure 2.9. The crossfeed conveyor and the surge deck in the Evans system.

Feeder built

Figure 2.9 shows the crossfeed conveyor and the surge deck. Figure 2.10 pictures the three-section feeder that Totem designed and built. A jumble of 8-, 12- or 36-foot-long logs with a single-section feeder would simultaneously feed double and triple logs onto that crossfeed conveyor. It just would not work. So the feeder was designed in three sections. Each section operates independently, or two or three can be operated together. There is also an ingenious hydraulic device to cut the size of that feeder in two to keep from double feeding small-diameter logs into the crossfeed conveyor. It has worked quite well with the large diameter variation—4 to 40 inches—that we have. The hydraulic device is needed to keep from double feeding but also makes it possible to kick into the oversized log pocket.

Figure 2.4 shows the feeder with the gate down and a log rejected and rolled out of the track. The system was designed for logs from 8 to 35 feet long and from 4 to 22 inches in diameter. During initial installation, logs were arriving that were up to 55

Figure 2.10. The three-section feeder designed and built by Totem for the Evans system.

Figure 2.11. A wide variation of log length and diameter causes a problem on the surge deck.

feet long. There is no way to put 55-foot logs into 40-foot pockets. Those logs had to be kicked out or they would have used two pockets.

Crossfeed conveyor

Note Figure 2.3 again. It shows the crossfeed conveyor coming onto the surge deck. Here logs were kicked off onto a 20-foot section. A problem on the surge deck turned out to be a short log against the feeder and next to it a 20- or 40-foot log. With live chains, the long log wanted to swing around and hit the limit switch and pretrigger the centering device, which would center the logs on the passing pair of chairs. It was necessary to put another set of chains in to lengthen this deck and install two pin stops in it to prevent these logs from going on through and triggering the centering device.

Figure 2.11 shows the problem. None of the Scandinavian systems we saw tried to handle this wide variation of length and diameter. The control house has to be moved back about five feet; otherwise the first log with a great deal of sweep or crook that went down the sort track would wipe it out.

Look at Figure 2.3 again for the microswitches and the surge deck. If the operator wants to add a log bucking program here, he is going to need the crossfeed conveyor. If sorting only is desired, about $60,000 of cost is eliminated; there is also a reduction in maintenance costs and in spots that may cause downtime.

Mechanical selector

The selector is geared directly to the idler sprocket on the end of the sort chain. A constant relationship is maintained between a set of sprockets and a roll of bars. When one key is pushed down simultaneously with the log rolling onto the sort conveyor, the log has already been centered on the chairs. The operator keypunches the species and diameter class. A simple, mechanical gear relationship continues down the sort line.

There are 15 rows of bars, about 24 degrees apart on a circle. There are two rows with 11 microswitches each. Pressing the key causes a cam action. As the cam travels around, it hits one of the

Figure 2.12. A view of the chairs looking down the sorting track from the feeder. The three concrete vents in the pockets created a problem area. A great deal of breakage occurred when logs dropped into the pockets.

22 microswitches. That gives a pulse to an electromagnet, tripping a finger to select the proper pocket.

This is log sorting with a strictly mechanical system. Figure 2.12 views the setup of the chairs looking down the sorting track from the feeder. The three concrete vents in the pockets, which can be seen in Figure 2.12, created a problem area. When logs dropped

Figure 2.13. Chairs slide down the rails in upright position. The log drops at about a 22-degree angle.

into the pockets, there was a great deal of breakage. (This has been corrected, as will be explained later.) The chairs that are sliding on the rails are held together by pieces of anchor-type chain. They are hooked fore and aft. An object getting into the way of the system and causing a malfunction will cause the chain to break and the hook to come off in such a way that it will not tear up the system. Figure 2.13 shows the way the chairs slide down these rails. The upright position is shown. The log drops at about a 22-degree angle. The hook (shown in the up position in the photograph) is caught and the rail is pulled up and reset.

Log sorter types

There are two types of arrangements for sorting logs. Logs 26 feet and shorter require a single-pocket release. In the Evans system, where logs are sorted to 36 feet long and 30 inches in diameter, there is a double-pocket release. A single-pocket release long enough to accommodate those chairs would create the problem of the chair trying to cock, as the rail is too heavy to be picked up and reset. Figure 2.14 shows how the chair tips when the log is dumped and rolls straight down into the concrete pockets. These pockets also caused considerable trouble; the contractor left the reinforcing steel out at the tops and the tops had to be replaced.

Figure 2.14. When the chair is tipped, the log is dumped and rolls straight down into the concrete pockets.

Breakage correction

To correct the problem of logs breaking when they dropped into the bunks, the stanchions were widened out and hog fuel was dumped in. The hog fuel helped arrest the rolling motion of the log. I have not seen any broken logs since the hog fuel was added. Some room was lost, but pocket capacity is still the 75,000 pounds that the stackers handle (Figure 2.15). We took into consideration from the beginning that breakage would be a problem. The price we had to pay was a little shallower pocket.

In summary

The Sateko sorting system allows logs to be sorted by species, grade, diameter and length in up to a maximum of 60 different pockets. It will allow a sawmill the following advantages:

1. An 8 to 15 percent better recovery from the log itself. The difference is in the ability to obtain additional boards instead of chips by properly programmed sawing techniques.

Figure 2.15. To arrest the rolling motion of the log and prevent breakage, stanchions were widened and hog fuel was dumped into the bunk. Although some room was lost, pocket capacity is still the 75,000 pounds that the stackers handle.

2. A minimum of a 25 percent increase in the capacity of the mill due to handling uniform-size logs.
3. Up to a third less labor because in a linear programmed mill sawing by diameters, you have less pieces.

DISCUSSION PERIOD

SANDY SANDERSON, International Paper Co., Veneta, Oregon: How long have you been operating this unit?

JOHNSON: We started serious debugging about April 5, 1972. Sateko's chief installer came over from Finland. We have been off and on the line for various reasons. The concrete failure cost us almost two months; we had to jackhammer the tops off and repour the cement. We have had other priority jobs that required us to leave the log sorter. All in all, we have had probably a month of steady running in the 10 months since April 1972—not much more.

SANDERSON: In that time, have you seen serious clean-up problems under the decks, transfer chains, loaders and so forth?

JOHNSON: That is one portion of the log sorter that has not been completed. We started on the waste system and got bogged down in the area where the logs dump onto the loading deck. Then we decided to get the rest of the system operating and worry about the waste system later.

VADNAIS: Cost of this system on a turnkey basis, as set up in Missoula, was $360,000. It is going to run about how much?

JOHNSON: Now we are at about $325,000, and we do have the waste system to go. But, hopefully, not much more debugging will be necessary.

VADNAIS: The system in Finland was designed for logs from 5 to 22 inches in diameter. Length depends upon your preference, but the pockets must be designed for the maximum length log you will have. In cooperation with Pacific Car & Foundry Co., we would prefer to design, engineer, build and erect these systems on a turnkey basis; we will turn them over to clients operating and running.

3

Electronic log inventory control systems

Edgar "Ed" McGrew, McGrew Bros. Sawmill Inc.
Ashland, Oregon

The McGrew Log Inventory Control System we shall discuss in this chapter is the result of a movement started in 1968 at McGrew Bros. Sawmill, Ashland, Oregon.

McGrew is a relatively small operation, cutting about 40 million board feet annually. In order to get our fair share of the market, we knew that we must work—not necessarily harder, but smarter. The Log Inventory Control System is one of the results of that effort. It is helping to make the company competitive, and it will help keep it that way.

For the first time there is a management information system for a single-plant or multiplant company that will print out on one man's desk whenever he wants it—every Friday or Monday morning, for example—a "snapshot" of his entire operation. This helps him make decisions to produce the best possible recovery of raw material dollars based on current end product market situations.

McGrew's Log Inventory Control System, which is called LOGINV, came about primarily because the company was using a tag system. We had long been aware that this system harbors continually growing error from one physical inventory to the next. This error accumulates because keypunch operators can input the wrong data, tags are lost, logs are broken, and so forth. Once a tag and its log become separated, there is no way to deduct this item from the inventory. As a result, we wind up with logs lost in a floating paper inventory. The only ways to adjust are to take a physical inventory or cut to bare ground.

Development of the LOGINV

The situation nets out thus: As we studied mill inventory operations around the country, we found that manual inventory methods resulted in inventories off as much as 10 percent or more. This error could be a company's operating margin. My father, Pink McGrew, and I decided that there had to be a better way. This was the beginning of the development of the concepts that ultimately led to LOGINV.

From the beginning, we looked at the problem from a viewpoint broader than just the contents of the log deck. We knew that if the right kind of information was available at the right time—

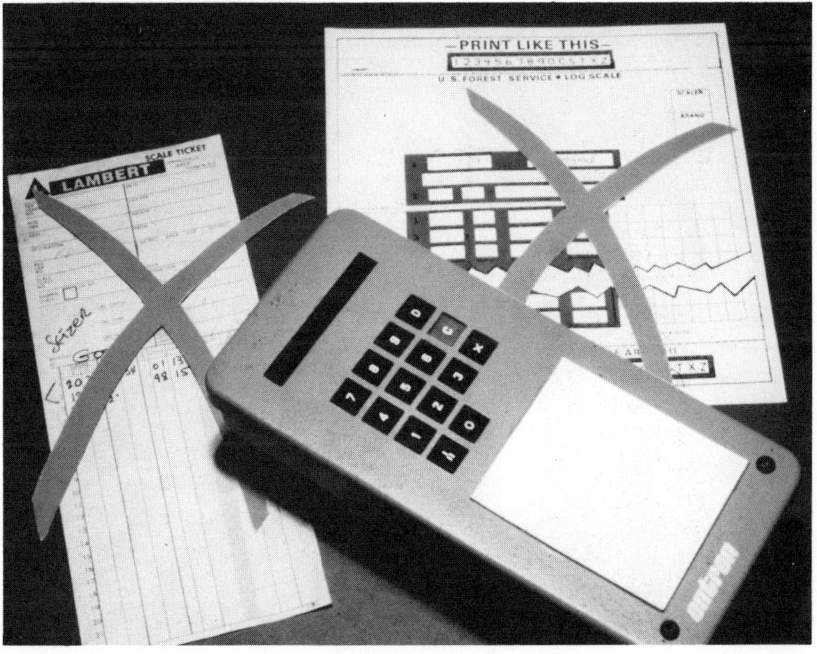

Scaler records data for computer via Entron solid-state electronic unit.

monthly, weekly or daily—we would have a powerful management tool that would assist in making more effective decisions. This concept led to the six reports we now have. Only one is a current inventory report.

One of the things we wanted to know was how well we were progressing on sales we were logging. The questions were, "What was the quality and delivered cost of the logs in the deck?" and "How true were the original cruise estimates to the final cut?" We wound up with the Area Report that answers these questions, and others. The source-type of information resulted in an opportunity for us as managers to evaluate how well we fared with a given sale

Truck scaling for the LOGINV program.

in a given area in a given time. Timber changes from one location to another. We could now tell, for example, that we could buy logs in a certain area; we could even pay a premium price. Of course, we would have much of this information by the seat of our pants, but now the automated system qualifies it and ties it to a volume and dollar figure. The Log Inventory Control System gives the true cost of getting a log into inventory. Such information is certainly a giant step ahead of the tag system, which only addresses the log in the yard.

The six reports in the new inventory system give a progressive picture of how well an operation is performing, how well it is

Scale information can also be obtained from deck or rollout.

doing in the woods, the yard and, to an extent, how well it is doing in the plant.

This system was developed on a time-shared computer. In other words, McGrew paid for just the time the computer was actually working on our project. After a couple of false starts, I did an intensive survey of computer power available in the Northwest and found that Computer Science Corp. was the only company offering a strong technical support group. This company has the largest network of computer power in the world with more than 6,000 employees. An extensive forest products library of programs is available—from timber resource allocation through cruising to paper production control. Computer Science has added another dimension to that library—the log inventory control program. It is now available from coast to coast and in a number of foreign countries. This log inventory control program provides a uniform information system anywhere a phone line can reach.

Data collection systems

It was apparent to us from the outset that both the best inventory control system in the world and the largest computer com-

Hard copy of scale ticket is created at the teletype at the scale shack.

plex had a severe limiting factor—data collection. Let us review two existing systems of data collection: (1) the old method and (2) the optical scanning method. Under the old method, the scaler measures and grades the logs and writes out a truck load ticket, eventually to be keypunched and, sooner or later, cranked into a computer memory. The scale information may become available up to 10 days later.

Significant changes could occur in the inventory between the time the log left the sale and the time the report arrived. The plant could have cut all those logs and the following half-week's logs. This is typical in the industry. Computer reports under the manual system are nothing more than pages from the history book of a sawmill.

There are other serious drawbacks to the manual method of data collection. The keypunch operator may not always read correctly what the scaler wrote, or the scaler may have written illegibly. Consequently, bad data are introduced into the system. Add to that what weather does to the scaler's paperwork in the field and the time spent copying it over into legible form, and you have a costly approach.

Optical scanning requires precise penmanship, keeping the marks inside the lines and writing perfectly. It is a slow process at best because scalers are not artists. Errors easily result from erasures. It is almost impossible to fill out one of those tickets when the temperature is 10 degrees and the snow is blowing. Even if scalers write the information correctly, the data go to the computer center at some later point in time. Sooner or later, you will get reports that, once again, are history.

Entron's data collection device

Obviously, we had to come up with a better data collection method. We looked toward electronics. To start, we surveyed tape recorders and data collection devices used for indoor inventory. They were not very portable and they had severe limitations. But we wanted to check to see whether we were on the right track. We scheduled a demonstration during August 1972 at S.H. & W. Lumber Co., Grants Pass, Oregon. We invited Forest Service people and sawmill owners in the area. They looked at what we had and they made recommendations. We used those ideas to design the basic unit in our new control system.

Built by Entron Inc., Richland, Washington, the data collection device is unique. It contains solid-state electronics, so it will last a lifetime. It is not heavy, weighing approximately 3 pounds. And it is compact, measuring 4 inches by 9 inches and less than 3 inches thick. It operates on a nickel cadmium rechargeable battery that gives a minimum of 8 hours use.

Because it is going to be used in all kinds of weather, the data collection unit operates from zero to 100 degrees Fahrenheit. You can drop it in the creek and it will still operate. It is dustproof and

Barker operator punches in data at the mill end of the log deck.

antimagnetic. Keys used to input the data are recessed so they cannot accidentally be activated and they are of a size that allows the scaler to wear gloves. When the keys are pressed, the information is displayed on a solid-state display that can be read in bright sunshine like the numbers from a pocket calculator. If the scaler makes an error on a log, he spots the error immediately, cancels it out and puts in the correct data. So in one operation he is your scaler, editor and keypunch operator.

The Entron data collection device takes advantage of the latest in electronic circuitry. It is a logic unit that was not even invented until just recently. It is an integrated-circuit, solid-state device used to store more than 1,000 characters of information in the memory of the Dataholder. The newest model has 8,000 to 10,000 characters of memory. It requires a secondary battery to power the expanded memory. This model can be used by cruisers and ground scalers.

Recently, a leading trade publication in the field of data processing reported a hearing involving Richard Case, director of advanced systems at International Business Machines. Mr. Case described the device in detail, calling it an "electronic notebook" that might be generally available some day. The fact that IBM thinks this a needed machine to speed up data collection gives the timber industry an opportunity to be a leader in the field of computer hardware and systems, truly at "the state of the art," as they say.

How the system works

With this background and explanation of the hardware, we would like to go through the procedure and explain how the system works. We begin by gathering data on each log. Initial data can be collected at the scaling ramp out in the woods or when the log arrives at the yard.

Final data are collected when the log leaves the deck and is sent to the barker. That gross scale remeasurement is necessary because the log must be deducted from the inventory to get an up-to-date record of the volume of logs on hand.

The initial scaler enters such information as grade and species, gross length and diameter. He also notes any defects. After entering the information, he hits the "D" key and the information is displayed. If it is correct, he dumps it into the machine's memory. If it is not right, he cancels and keys in the correct information.

When the scaler finishes with the load, he steps into his shack and slips the Entron device into what we call a "black box." He does not connect plugs, wires or anything. The connection is made automatically. This unit or formatter, as it is known, is connected to a teletype machine standing beside it.

When the scaler hits a button, all the information held in the memory of the Dataholder is dumped into the formatter, which, in turn, directs the teletype to print out a scale ticket for that load. The driver gets a copy for his record and heads to the deck to unload. When the teletype prints out the hard-copy record, it also records the information on paper tape or on a magnetic tape cassette. At the end of his shift, the scaler takes the tape, along with the office copy of all loads, to the office. Verification occurs when the data are transmitted to the computer over the telephone line.

Input from the barker

As noted above, the log is measured just before it moves from the deck to the barker. At that time, only the gross measurement, length and diameter, plus species and grade, are taken. The men here are not expert scalers, so we do not ask them to figure defects. The computer has already figured the average defect for each grade in each species, and it automatically subtracts the average deduction for each log when the scaling information is fed into the computer at the end of the day. The test volume came out to within about 4 percent of the original scaler. That is closer than when two men judge the same log. The system allows machine operators to record plant usage.

At the end of the day, all data collected at the barker are fed into the computer. The process is so simple an office clerk can be taught everything she or he needs to know in a few hours, including getting the reports that are printed on the teletype machine out of the computer. For example, to get a specific report, the clerk merely types in the word "RUN" and the name of the program. The clerk has to learn a few commands, learn a few elementary things about editing on the machine, and that is it. The computer now has all the information it needs to pump out any of the six reports for management evaluation.

Management's six reports

The key report among the six reports in the inventory system is the Current Inventory Report that shows the actual log inventory as of the date the last load was received. It quickly points out production run opportunities. Alternatives for production or purchasing can be evaluated quickly, and decking becomes a more precise action. Combine this month's report with the one from last month, plus other reports, and the result is a close look at the inventory value versus production performance.

The Trucker Report gives a breakdown of receipts by each trucker. It gives details of each load to date, the sale area, who the trucker is, the scale ticket number and the number of logs for each load plus gross and net footage. The report helps evaluate perform-

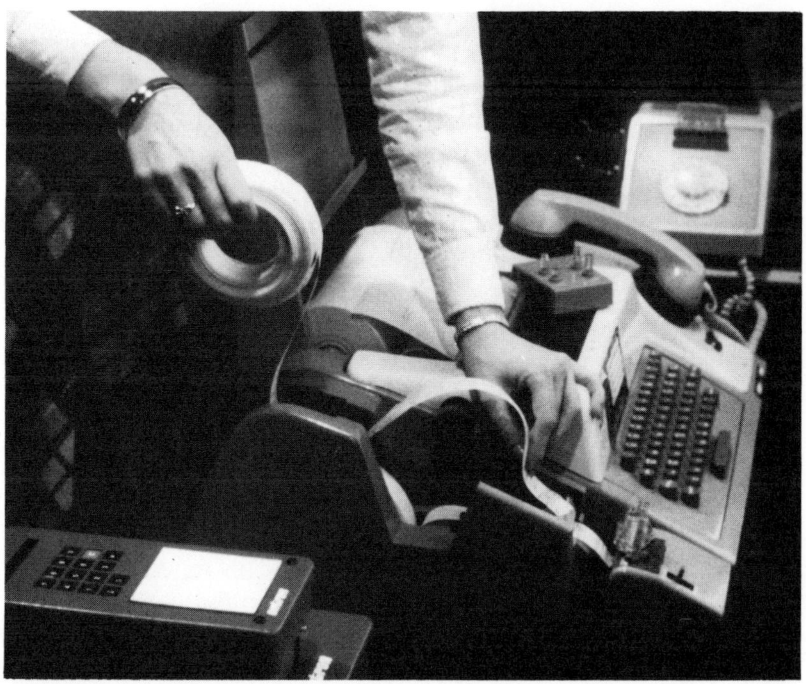

At day's end, office employee transmits transactions to the computer.

ance of drivers. It spots a man who is not bringing in the volume. Maybe he is spending a couple of hours down at the cafe making time with a waitress, or maybe his truck is breaking down all the time. Either way, it is costing money. The report gives an indication of where the trouble is. The report also provides basic information necessary to calculate the pay for gyppo drivers hauling by the thousand. Here, too, is statistical information pertaining to log hauling cost.

The Area Report is especially valuable. It shows the results of buying from a particular source. The volume of the grades in each species and the percentage of deduction for each one are quickly seen. If needed, the report can identify the number of logs in each diameter within each grade and species. The source report is a tool for quantitative evaluation of timber sources in terms of production, quality and cost. Most important, it gives management the critical information needed to guide future buying effectively. On the left side of the report is the information for the current report period. On the right side of the report are the totals since the sale area was opened.

The Master Source Report takes all the individual Area Reports and consolidates them to indicate to management the quality and acquisition cost of timber per thousand board feet from all suppliers. With this management tool, a forecast can be made on buying needs based on current mill production and new inventory added. The profitability of the buy based on market conditions can be targeted, and future activities can be adjusted accordingly.

If desired, a daily usage report can be run to check the day-by-day operation of the mill. This gives information concerning mill input by shift, volumes by grade and species and raw material costs associated with production.

Finally, going one step further, the Period Usage Report combines all the daily reports during the current period. This report gives management a broad insight into the overall performance of the operation. One glance will identify raw material costs associated with production for the entire period. This report, compared with tally information, points up the quality and quantity of recovery from raw material. Using it with current market information will give an assessment of the marketability of the dimensions and species produced.

Special value to integrated mills

The Log Inventory Control System is most attractive to integrated mills operating under one management where there is, for example, a veneer plant and a sawmill. The manager can consider the market for each product, compare it with the inventory and make a judgment as to the most profitable product to produce. For example, a certain type of log like a peeler is worth more today cut into a vertical grain clear or shop lumber than it is into plywood. If a manager knows his inventory and the volume of each type of log, he can play "what if" games with his computer and schedule a run that will bring him the biggest return at the time based on the current market.

New developments ahead

Now, naturally, we are not at a dead end. There are new things coming. We will soon introduce a new series of reports that deal with the lumber inventory through dry kiln, the planer and out through shipping. This analysis will yield a number of records for accounting, including identifying dollar amounts received for each product as it went down the track.

Let me give an example of the information this new report will deal with. It will allow comparison of the production of a sawyer on day shift with the night-shift sawyer. We ran this program and found that one sawyer was not producing the same quality of lumber as the other sawyer even though he was sawing essentially the same logs from the same sale. We sat him down, showed him the report, motivated the man and then kept a close watch on his work. Almost immediately, we realized a gain of about $200 profit from his shift. That is bottom line money—$1,000 a week, $50,000 a year—that could have been lost if a mill were not keeping close tabs on production and costs.

Costs detail

Now what does such a Log Inventory Control System cost? Our survey of mills in the 30- to 40-million-board-foot category shows that it costs about 4½ cents a log, covering the handling of the log when it comes in and again when it heads for the saw. Let us compare this to other systems.

One company about the size of McGrew reports that its system costs 7 1/3 cents a log, and they do not get half the management information.

For the first time, we have a cost-effective management information system—good for a single-plant operation or a multiplant company—that will give management timely, easy-to-understand reports to help make decisions that will result in maximum profit for the company. This is what it is all about. When the industry tightens up, as it always does, the firms with the best management practices will be the ones that stay on top. We believe our Log Inventory Control System will help us do just that.

DISCUSSION PERIOD

UNIDENTIFIED: What is the cost of the data recording unit? And what does it cost to get the information back out if you do not have a computer system?

McGREW: Without a computer system, I do not know how the unit would work. Cost of the Dataholder in the 1K memory unit is about $1,595. The formatter, which the dataholder drops into and which instructs the teletype or other terminal, is about $2,400. That is single-unit quantities. So far as doing without a computer, I really do not know.

GUS HALEY, M&R Timber Co., Port Angeles, Washington: Are you using all of your own logs? Do you sell any of your log inventory?

McGREW: We are buying logs from Weyerhaeuser and at government sales. The only inventory we sell is cedar logs.

HALEY: How do you handle the inventory to be sold on scale-out?

McGREW: The cedar that we sell is scaled by the yardman, just as if it were a truck coming in. Then we put a minus sign behind the end of the entry header. This backs it out of inventory.

4

Best opening face for second growth timber

Hiram Hallock, Technologist, Division of Wood Quality
Research, U.S. Forest Products Laboratory
Madison, Wisconsin

Work on the mathematical relation of sawing factors such as kerf, oversizing and lumber sizes to yield from small logs was begun at the U.S. Forest Products Laboratory in Madison, Wisconsin, more than 10 years ago. We believed there must be a better solution to the problem of yield maximization than the diagrammatic approach then generally in use.

In our earliest approach, instead of analyzing what a certain log would yield, we started from the other end and mathematically assembled the various combinations of dimensions and boards. Then we calculated the size of the circumscribing circle. Thus we were able to determine the critical change points where size differences in kerf or lumber cause yield changes.

Importance of kerf width

A theoretical study was conducted, with results published in 1962. This study effectively demolished the much-repeated statement that changes in kerf width or lumber thickness have no effect on yield unless the log is large enough to yield an extra board.

I was recently encouraged when Carl Mason, a Gladstone, Oregon sawmill engineer, told me that he had carried the published report in his briefcase for years in his efforts to convince men in the sawmilling industry of the importance of the reduction of kerf width and the control of lumber sizing.

Converting more of the log to lumber

More recently, in response to a deepening supply-and-demand crisis in the national softwood situation, we began to search for better ways to make lumber-type products. Our basic objective was to convert more of the log to lumber. One of the better ideas we have devised is a system known as EGAR, which means edge, glue and rip. I will not explain the system in detail now except to note that spin-off from this theoretical work led to our Best Open-

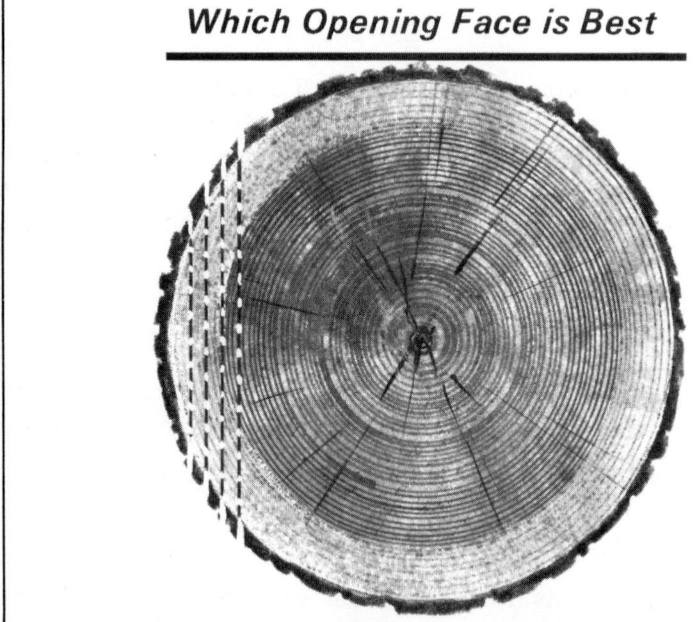

Figure 4.1. Location of the slabbing or opening face on both log and cant is the key to the sawing pattern that maximizes yield, because the first face establishes the position of all other faces.

ing Face (B.O.F.) program, originally a highly versatile mathematical model of the sawing process.

We at the Madison laboratory are not the only people working on lumber yield maximization. There are a few others engaged in this effort and some of their work is very good. As the idea of yield maximization through process control gains momentum in the industry, we expect much more to be done. Yield maximization helps alleviate the timber shortage; at the same time, it converts logs into lumber more efficiently.

If logs could be grown as rectangular solids instead of approximate conic sections, their conversion to lumber could be more efficient. Basically, sawing is fitting rectangular solids (lumber) into somewhat cylindrical logs or, more accurately, extracting rectangular solids from cylindrical logs. In all sawing, location of the slabbing or opening face on both the log and the cant is the key to the sawing pattern that maximizes yield, since the position of the first face establishes the position of all other faces (Figure 4.1).

A given set of sawing conditions such as kerf width, sawing variation and shrinkage established the thicknesses and widths of

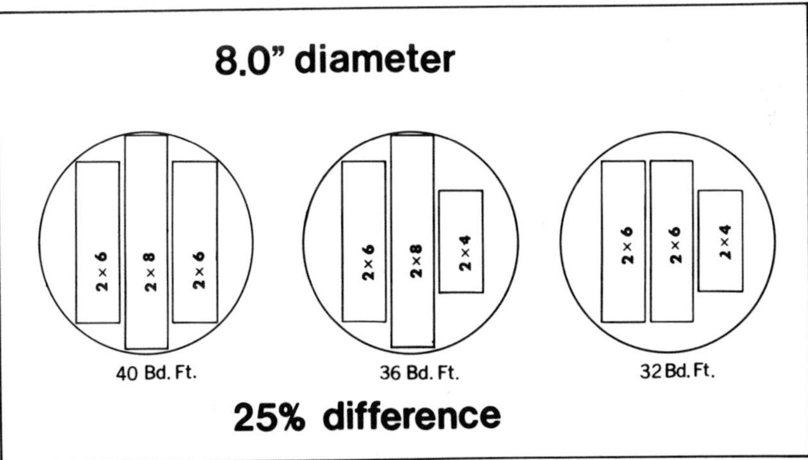

Figure 4.2. An example of the effect on lumber yield when saw lines are shifted back and forth across the face of the log ends. An 8-inch log, 12 feet long, is sawn into nominal 2-inch dimension. Although the three solutions look similar, the cut on the left yields 40 board feet, a 25 percent greater yield than the cut on the far right.

rough green lumber. Each lumber item has a definite scale. Total yield is the sum of those items that can be fitted into the log diameter being examined. Shifting saw lines back and forth across the face of the log end results, in almost all cases, in substantial differences in lumber yield due to the critical widths of green lumber blanks and their geometrical interrelation for all different log diameters.

Figure 4.2 is an oversimplified example. Here is an 8-inch log sawn into nominal 2-inch dimension—no boards. Let us assume there is no wane allowed on the lumber. The same geometrical factors are involved whether we allow wane or not, but it is easier to illustrate without rounded wane corners. To make yield calculations easy, let us assume the log in Figure 4.2 and the logs in the two figures to follow are 12 feet long.

Best cut patterns

At first glance, the three solutions presented in Figure 4.2 look similar. Actually, the cut shown on the left is the best solution

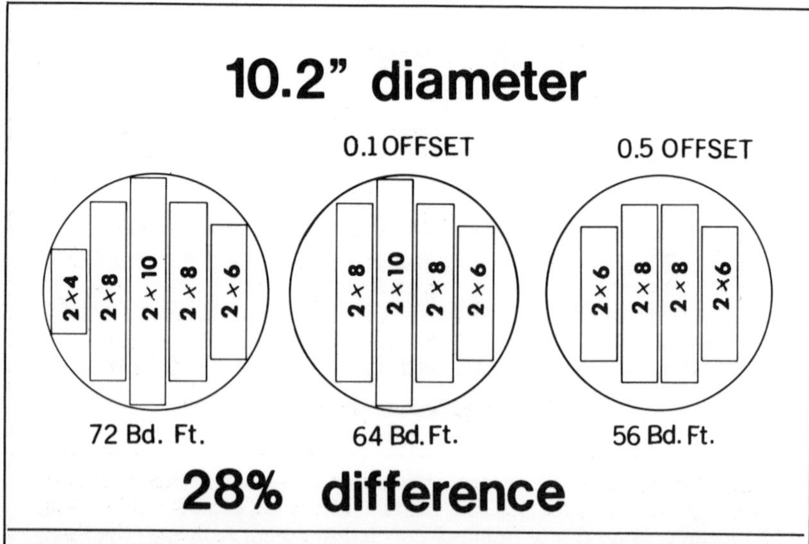

Figure 4.3. With a 10.2-inch log, 12 feet long, the best sawing solution yields 72 board feet, 28 percent more lumber than the poorest.

with a yield of 40 board feet. When the saw lines on the log are shifted only 1/10 inch to the right (center), the yield drops to 36 board feet. At right, the saw lines are shifted 2/10 inch to the right and the log yields only 32 board feet. The best solution is 25 percent better than the poorest.

Figure 4.3 shows a 10.2-inch log. The illustration at the left in Figure 4.3 shows the best solution yielding 72 board feet. If the sawing pattern is shifted only 1/10 inch to the left, the 2x4 on the left is lost and the yield drops to 64 board feet. A saw line shift of 1/2 inch to the left results in the lowest yield of only 56 board feet. There is a 28 percent difference between the best and the poorest solutions.

Figure 4.4 shows a 13.2-inch log. Again the cut on the left is the best solution with a yield of 128 board feet. Moving the saw lines 1/10 inch to the left drops the yield to 124 board feet. If the pattern is shifted 2/10 inch to the left, a low yield of 112 board feet is obtained. In this case, the best solution is 14 percent better than the poorest.

This is an oversimplification, but there are good sawing solu-

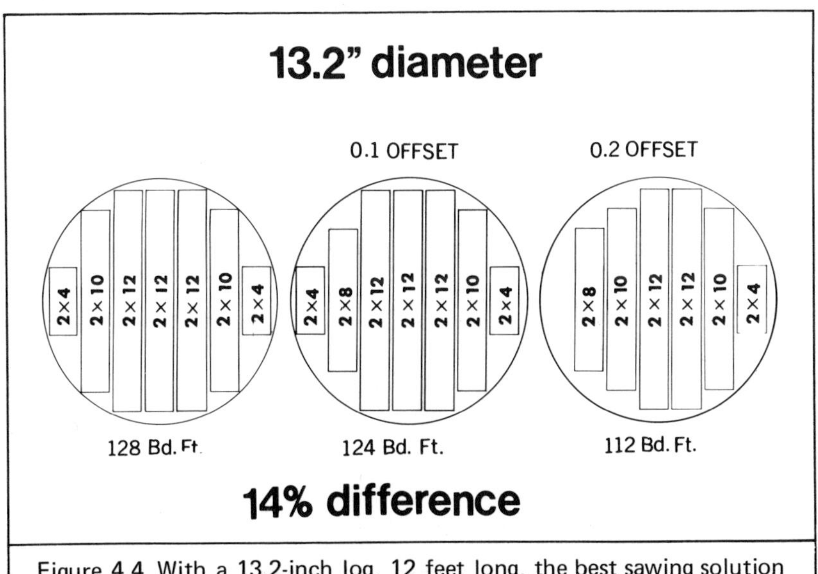

Figure 4.4. With a 13.2-inch log, 12 feet long, the best sawing solution yields 128 board feet, 14 percent more lumber than the poorest.

tions and poor sawing solutions. The good sawing solutions are appreciably better than the poor ones.

The three illustrations involve simple live sawing where all saw lines are in the same plane. When cants are produced and the cants are then sawn, there is a second sawing plane and the cants also have their good and poor sawing solutions.

Problems with use of diagrams

Attempts to obtain maximum-yield solutions through construction of diagrams have been made. The results were undoubtedly better than hit-or-miss sawing. However, the number of possible saw-line combinations in the production of side lumber and a cant multiplied by the many possible saw-line positions in the cant itself for each log diameter and length are astronomical in number and so preclude the examination of more than a small percentage of the possible solutions. And if only one factor in the sawing process is changed—kerf width, for example—you have an expensive set of absolutely worthless diagrams.

We decided, therefore, to modify our mathematical model of the sawing process so that, for any specified set of sawing conditions, it would be possible to look at all sawing solutions for each log diameter and determine the best solution in each individual case.

Program Variables

1. Thickness of lumber dry
2. Width of lumber dry
3. Planing allowance per face
4. Shrinkage during drying
5. Sawing variation (scant)
6. Saw kerf
7. Log diameter range
8. Log diameter increment
9. Face opening increment

Figure 4.5. Variables in the initial version of the B.O.F. program.

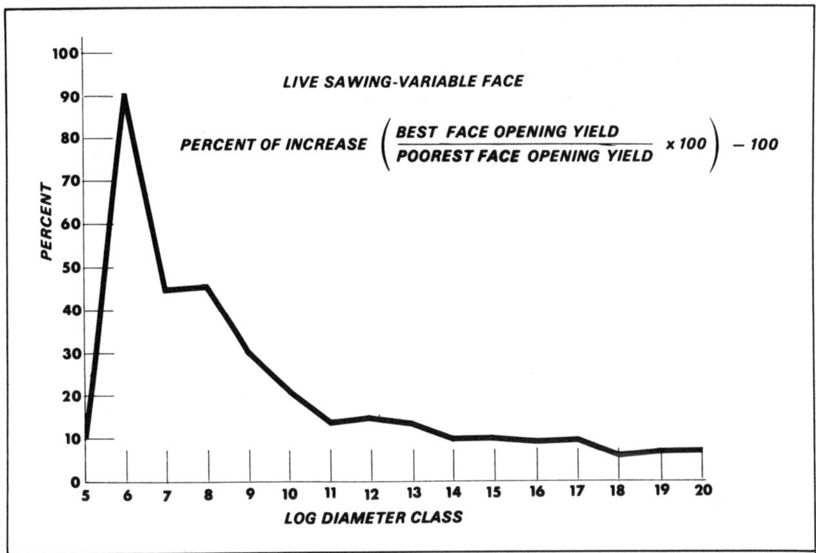

Figure 4.6. The difference between the best and poorest solutions for live sawing in the 5- to 20-inch diameter range averages 21 percent.

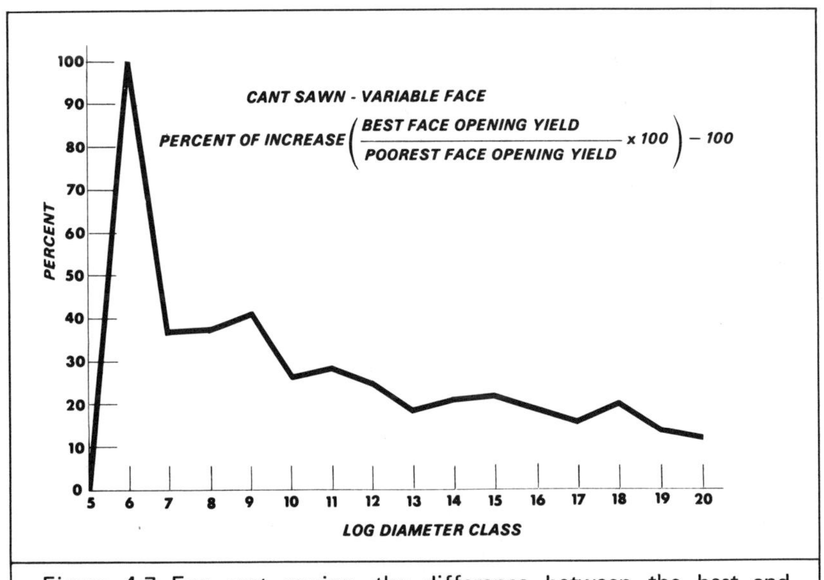

Figure 4.7. For cant sawing, the difference between the best and poorest solutions averages about 27 percent.

Initial version of B.O.F.

The initial version of our B.O.F. program considered dry finished lumber sizes, planing allowance, shrinkage, sawing variation, saw kerf width, log-diameter range, diameter measuring increment and incremental capability of the sawing equipment to produce a given opening face (Figure 4.5).

The results obtained from analyzing the actual difference between the best and the poorest solutions of the sawing problems surprised us. Figure 4.6 reveals what we found for live sawing logs in the diameter range of 5 to 20 inches. The line on the chart shows the percentage by which the best solution exceeds the poorest. The difference ranges from 6 to 90 percent; it averages across the board about 21 percent. The difference is even more spectacular when cant sawing is used (Figure 4.7). The best solutions exceed the poorest by 12 to 100 percent and average about 27 percent.

In both cant and live sawing, the differences tend to decline with the larger diameters. This is logical since the fitting problem is reduced. In poor solutions, the problem is that several flitches are produced that are almost, but not quite, wide enough for an

---

B.O.F. - COMMERCIAL VERSION

Added Variables	User Options
log taper	board & dimension or dimension only
resaw kerf	live or cant sawing
narrowest piece	volume or value maximization
shortest piece	green or dry ALS sizes
lumber prices	lumber mix bias

Figure 4.8. Variables and user options added to the commercial version of the Best Opening Face program.

---

additional 2 inches of nominal width. This problem represents a smaller percentage of the whole as the log diameter increases.

Maximum yield study

We published the results of our maximum-yield study late in 1971. Response from the sawmill industry bordered on the unbelievable until I attended the Sawmill Clinic. Then it became believable. We have answered industry requests for about 1,000 copies of the original publication, in addition to requests from the general public and the Forest Products Laboratory's regular mailing list. I think this is a favorable and powerful comment on industry's consciousness of the need to improve lumber recovery.

When we say lumber recovery factor (L.R.F.), we mean board feet of lumber per cubic foot of log volume. Studies in both the United States and Canada indicate that mills get lumber recovery factors ranging from below 4 to a little above 9 board feet per cubic foot. The industry average is probably about 6½ board feet.

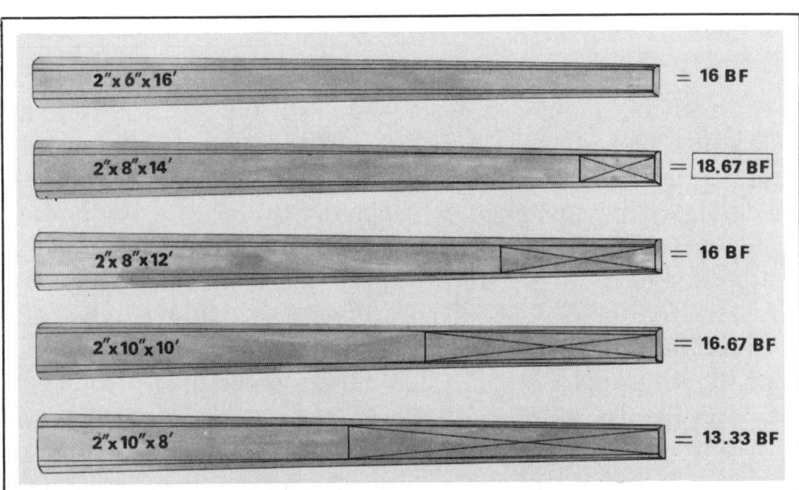

Figure 4.9. The maximization routine for edging and trimming is one of the reasons B.O.F. finds maximum-yield solutions appreciably higher than industry averages. Here the best solution, a 14-foot 2x8, yields an additional 2 2/3 board feet or 16.7 percent more than the normal full-length edging.

As one might expect, the lumber recovery factor tends to be lower with small logs.

The research version of the B.O.F. program has proven that mathematically optimizing the geometry of sawing does indeed promise substantial gains in lumber recovery. For simplification and to save time, this original version of the program did not include some variables and options necessary for direct sawmill application. A program including these variables and options, in addition to those already referred to, is now complete, operational and, I am happy to say, in use.

Variables considered

In addition to the variables in the original B.O.F. program, the commercial version (Figure 4.8) now considers:

1. Amount of taper in the log.
2. Width of the resaw kerf.
3. Narrowest and shortest acceptable lumber.
4. Price of all lumber items of all lengths, when dollar maximization is desired.

Wane on the lumber in accordance with the National Grading Rule is also allowed.

The program user has the choice of 1-inch jacket boards and 2-inch dimension combined or 2-inch dimension only; live or cant sawing method; volume or value maximization; and lumber manufactured to green or dry American Lumber Standards (ALS) lumber sizes (Figure 4.8). Another feature allows the user to bias the program so that rather than truly maximizing either volume or value, it will produce a controlled mix of lumber items—to match the order file, for example—and, within this constraint, will maximize output.

When used as a processing decision tool, this program accepts mill conditions as they are and for each log diameter and length class determines the precise position of the opening face or faces that will result in either maximum yield or maximum dollar value. It determines the total lumber tally and, if dollar maximizing, calculates the product value. In addition, the lumber recovery factor is determined for each solution.

Another of the secrets—if that is the proper word—of this program's ability to find maximum-yield solutions appreciably higher than industry averages is its maximization routine for edging and trimming. Figure 4.9 shows a flitch that if edged full length will yield a 2x6 16 feet long. The program tries trimming to successive 2-foot shorter increments and determines at each increment the proper edging width within the wane rule limitations. It then selects the length and width combination that gives the highest volume or, if dollar maximizing, the highest value. Here, the best solution, a 14-foot 2x8, is 2 2/3 board feet or 16.7 percent more than the normal full-length edging would yield.

For flitches with a width at any point greater than required for a 2x12, another routine is followed. Figure 4.10 shows a flitch that if edged full length would yield a 2x12 16 feet long scaling 32 board feet. However, at the 10-foot point the flitch is wide enough to yield a 2x4 and a 2x10. Thus this solution—a 10-foot 2x4 and a 16-foot 2x10—has a combined volume of 33 1/3 board feet, or a gain of 4.2 percent.

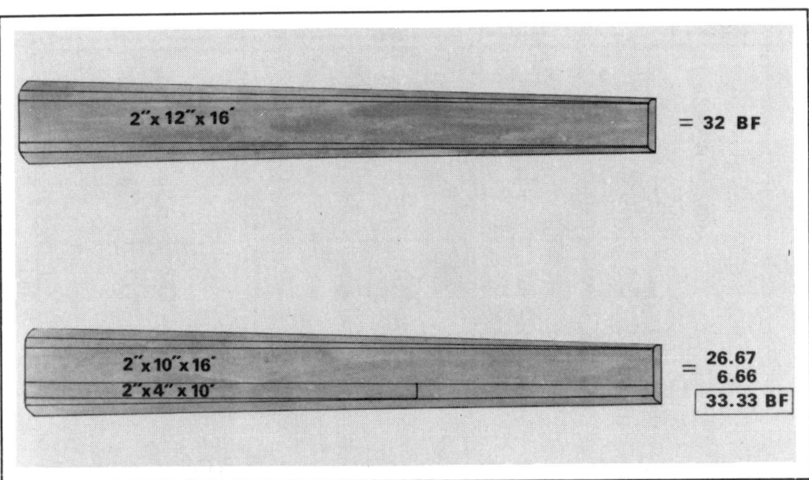

Figure 4.10. If edged full length, this flitch would yield a 2x12, 16 feet long, scaling 32 board feet. However, at the 10-foot point the flitch is wide enough to yield a 10-foot 2x4 and a 16-foot 2x10. Combined volume of the 2x4 and the 2x10 is 33 1/3 board feet, a 4.2 percent gain in yield.

Installation cost

Although we have revised our original installation cost estimates slightly upward, our information is that this system can be completely installed in a new mill for an additional cost not to exceed $125,000. An estimate given earlier in the Clinic was considerably lower than that. There exists no doubt in our minds, nor in those of many progressive people in the industry, that additional yields will be at least 10 percent when maximizing volume, and probably

Thick and Thin Lumber Company
Original Mill — Live Sawing

Log Diam.	Lumber yield	L. R. F.
8.1	53	7.3
13.1	139	8.0
18.1	264	8.3

Average L. R. F. 7.9

Average Overrun 34%

Figure 4.11. Thick and Thin Lumber Company's recovery in terms of board feet and lumber recovery factors with live sawing in the original mill operation.

more when maximizing dollars. Payout time for a mill cutting only 50,000 board feet a day, 12 1/2 million annually (and that is a small mill), should be about a year. Real payoff, of course, is that with no other improvements in the mill the same amount of lumber can be cut from 10 percent less cubic log volume.

In actual use, the B.O.F. program is run on a medium-sized computer using the sawmill's values for the variables and the user's options. This run gives the answers to saw-line positioning for every combination of log diameter and length. This information is then stored in the sawing system process control minicomputer. A log about to be sawn would be scanned for diameter and length and this data relayed to the minicomputer that signals the numerically controlled log positioner. The log, when properly positioned (that is, shifted horizontally as required), will then be sawn in the normal manner.

B.O.F. is versatile tool

Not only does B.O.F. solve the maximization problem but, because it accepts all of sawing's variable factors, it is also capable of answering all the questions in the "what is the effect of . . .?" series—such as, what is the effect of changing the headrig kerf from 1/4 inch to 3/16 inch? Or, what is the effect of reducing sawing variation by 1/8 inch? To illustrate this very useful side of B.O.F. in evaluating potential yield and lumber recovery factor changes in log conversion, let us assume the following milling operation, one that is reasonably typical of a still-too-large segment of the U.S. milling industry.

The Thick and Thin Lumber Company is sawing only dimension lumber on a circular mill with head saw and edger kerfs of 3/8 inch. The lumber thickness variation limit necessary to include 95 percent of their lumber is plus or minus ¼ inch. The firm is live sawing the logs and trying to maximize volume yield. The smallest lumber saved is an 8-foot 2x4.

Lumber recovery factors

Let us proceed through a series of changes in mill equipment and practice, looking at the improvement for which each is respon-

Thick and Thin Lumber Company
Band Headrig Added

Log Diam.	Lumber yield		L. R. F.	
	Before	After	Before	After
8.1	53	59	7.3	8.0
13.1	139	153	8.0	8.8
18.1	264	295	8.3	9.3

Average L. R. F.	Before	7.9
	After	8.7
Average Overrun	Before	34%
	After	49%

Figure 4.12. Thick and Thin's recovery after owner Smith installed a new headrig with a 1/8-inch kerf.

sible. One thing to be kept in mind is that the use of B.O.F. sawing solutions is assumed in all cases. This does not affect the validity or size of the indicated change, but it does make the yield levels at least 10 percent higher than most in the industry. To keep the data simple, let us consider that all logs are 16 feet long and of three diameters—8.1, 13.1, and 18.1. There is nothing magic about those numbers; I simply picked some diameters.

Figure 4.11 shows Thick and Thin's current recovery in terms of board feet and lumber recovery factors. The lumber recovery factors look high because the recoveries assume best opening face, as I said, and correct edging and trimming—something never accomplished by human operators. The average lumber recovery factor is 7.9; expressed in more common mill operators' terms, this is a 34 percent overrun over Scribner Decimal C. This is on short log scale, not long log scale (Figure 4.11).

Thick and Thin Lumber Company
Sawing Variation Reduced

Log Diam.	Lumber yield		L. R. F.	
	Before	After	Before	After
8.1	59	65	8.0	8.9
13.1	153	175	8.8	10.1
18.1	295	335	9.3	10.5

Average L. R. F.	Before	8.7
	After	9.8
Average Overrun	Before	49%
	After	69%

Figure 4.13. A new carriage and setworks of the digital, minicomputer-controlled type reduce sawing variation and increase lumber recovery.

New equipment improvements

Owner John Smith is convinced that replacement of his circular headrig will increase his lumber recovery factor. After installing the new headrig, which has a 1/8-inch kerf, his new recovery is shown in Figure 4.12. Note that even the small 8.1-inch log yields 6 board feet more with the narrower kerf. Total lumber yield is up more than 11 percent, which is reflected in the increased lumber recovery factor and overrun values. Average L.R.F. is now 8.7 and average overrun is 49 percent.

Owner Smith is so pleased with the increased yield that he makes a decision to go all the way toward maximum processing efficiency. Finances prevent his doing everything at one time, so he proceeds in an orderly, step-by-step fashion. A careful look at the operation indicates that his sawing variation of plus/minus 1/4

Log Diam.	Lumber yield		L. R. F.	
	Before	After	Before	After
8.1	65	65	8.9	8.9
13.1	175	177	10.1	10.2
.18.1	335	339	10.5	10.6

Precision Lumber Products, Inc.
Band Resaw Added — Cant Sawing

Average L. R. F. — Before 9.8 / After 9.9

Average Overrun — Before 69% / After 71%

Figure 4.14. Thick and Thin, renamed Precision Lumber Products, Inc., adds a linebar resaw with a 1/8-inch kerf, making it practical to change from live to cant sawing. Although the most important improvement is in increased production potential, yield is increased by 1 percent.

inch is, of course, much too high. His next purchase is a new carriage and setworks of the digital, minicomputer-controlled type. This equipment has a very accurate overall setting capability with no cumulative error. Careful checking of the lumber now shows that the combination of the narrow kerf band and the new carriage has reduced lumber thickness variation to about plus/minus 1/64 inch.

Now the lumber recovery picture has improved again. Lumber recovery is up 13.4 percent, the recovery factor is approaching 10, and overrun is nearly 70 percent (Figure 4.13).

To take care of the increased yield coming from a reasonably constant log supply, Smith's next improvement is to add a linebar resaw—also with a 1/8-inch kerf. This machine makes it practical

Precision Lumber Products, Inc.
Precision Lumber Products, Inc.
Planing Allowance Reduced

Log Diam.	Lumber Recovery		L. R. F.	
	Before	After	Before	After
8.1	65	67	8.9	9.1
13.1	177	187	10.2	10.7
18.1	339	351	10.6	11.0

Average L. R. F. — Before 9.9 / After 10.3

Average Overrun — Before 71% / After 78%

Figure 4.15. In a test run, Precision Lumber Products finds that it can now reduce its planing allowance from 2/32 to 1/32 inch per face. Reprogramming the minicomputer-controlled setworks to a 2/32-inch-smaller target lumber size increases lumber yield more than 4 percent.

to change from live to cant sawing. Figure 4.14 shows how this change affected the lumber recovery. Note that Smith was so proud of the improved manufacturing quality that he changed the company's name from Thick and Thin Lumber Company to Precision Lumber Products. The most important improvement this time, of course, is in increased production potential. However, yield was increased by 1 percent as a result of going to cant sawing instead of live sawing.

Further careful checking by Smith indicates that his new precision equipment has resulted not only in more accurate lumber, but in lumber with much less roughness. A test run shows he can reduce his planing allowance from 2/32 inch per face to 1/32 inch. This allows him to reprogram his minicomputer-controlled set-

Precision Lumber Products, Inc.
1″ Board = 3″ x 6′ salvaged

Log Diam.	Lumber Recovery		L. R. F.	
	Before	After	Before	After
8.1	67	73	9.1	10.0
13.1	187	192	10.7	11.0
18.1	351	367	11.0	11.5

Average L. R. F.   Before 10.3   After 10.8

Average Overrun   Before 78%   After 86%

Figure 4.16. Since Smith is recovering only dimension lumber 2 by 4 by 8 feet and larger, he decides to cut 1-inch jacket boards from those logs and cants where it is advantageous. He will also salvage all lumber 3 inches by 6 feet and larger. Now yield increases another 4.5 percent.

works to a 2/32-inch-smaller target lumber size. Recovery figures following this change show that lumber yields are up over 4 percent. The lumber recovery factor now exceeds 10, and overrun is approaching 80 percent (Figure 4.15).

Smith has now almost run out of improvements to make in his operation. The machining process is nearly perfect. However, he is still recovering only dimension lumber 2 by 4 by 8 feet and larger. He decides to cut 1-inch jacket boards from those logs and cants where it is advantageous. Also, he will salvage all lumber 3 inches by 6 feet and larger. Figure 4.16 shows what this change does to improve the recovery picture. Yield is up another 4.5 percent with

Precision Lumber Products, Inc.
Summary
Past and Present

Log Diam.	Lumber Sawing		L. R. F.	
	Past	Present	Past	Present
8.1	53	73	7.3	10.0
13.1	139	192	8.0	11.0
18.1	264	367	8.3	11.5

Average L. R. F.   Past __7.9__   Improvement __37%__
   Present __10.8__

Average Overrun   Past __34__   Improvement __153%__
   Present __86%__

Figure 4.17. Comparing present yield with figures prior to improvements, Smith finds board-foot yield for comparable logs up almost 39 percent. The lumber recovery factor has increased by 37 percent and the overrun shows an improvement of 153 percent.

the recovery factor approaching 11 and overrun more than 85 percent.

Increases in yield

It has now been a couple of years since John Smith made the first improvement for his old Thick and Thin Lumber Company. He digs out the old yield records and compares them with his latest yield records (Figure 4.17). Even he is surprised at what this comparison shows. Actual board foot yield for comparable logs is up almost 39 percent, the lumber recovery factor has increased from 7.9 to 10.8 (an improvement of 37 percent) and the overrun has risen from 34 to 86 percent. This is an increase in the overrun of 153 percent.

Perhaps most important to John Smith is that Brinks now makes two stops per day instead of one to transfer profits to the bank! That statement may be a little facetious—I have been in the sawmilling business, too—but there is nothing facetious about the increase in yield possible by making the changes just reviewed. The results are hard, cold mathematics. They are not subject to human bias. Failure to accept the general conclusion of analyses of this type is simply the approach of an ostrich with his head in the sand.

Diameter vs Yield

Log Diam.	Yield
8.1	74
8.2	74
8.3	76
13.1	192
13.2	194
13.3	196
18.1	367
18.2	373
18.3	374

Figure 4.18. Some yields for close increments of diameter.

Effect of Taper

Log Diam.	Taper/16′		
	2″	3″	4″
8.1	74	78	82
13.1	192	197	206
18.1	367	372	384

Figure 4.19. Yield increases significantly with increasing taper. There appears to be an inverse relationship to the diameter.

Green vs Dry ALS Size

Log Diam.	Yield		
	Dry	Green	% Increase
8.1	74	75	1.4
13.1	192	196	2.1
18.1	367	390	6.3

Figure 4.20. Yield increases are related positively to diameter and favor the green operation by 1.4 to 6.3 percent.

Here is the ultimate log scaling system—at least in our board-foot-oriented industry. When values for the mill variables have been chosen, the program can output yield table data for all increments of diameter and all degrees of taper. And these two items are important. Currently, scaling is usually by 1-inch classes and average taper, and this approach covers up substantial differences by ignoring the actual diameter and the actual taper.

Taper affects yield

Figure 4.18 shows some yields for close increments of diameter. Note that differences are real between 1/10-inch classes, especially in the larger diameters.

Log taper turns out to be a real surprise with yield increasing very significantly with increasing taper (Figure 4.19). It also appears to be inversely related to the diameter.

Finally, scale should perhaps be different if the mill cuts to green lumber ALS sizes rather than dry. Figure 4.20 shows increases related positively to diameter and favoring the green operation by 1.4 to 6.3 percent.

Figure 4.21 compares results of volume maximization and value maximization for the three log diameters we have considered.

Volume vs Value Maximization

Log Diam.	Volume Max.		Value Max.	
	Volume	Value	Volume	Value
8.1	74	11.73	74	11.73
13.1	192	27.94	189	31.02
18.1	367	$60.57	366	60.67

Figure 4.21. A comparison of results of volume and value maximization for the three log diameters considered.

Note that for the 8.1-inch log, volume maximization and value maximization have the same solution. For the 13.1-inch log, value maximization obtains three less board feet but its lumber mix is worth \$3.08 more. This is a very substantial difference. Little difference exists in the 18.1-inch solution, with value maximization yielding an additional value of 10 cents, but one board foot less.

Availability of the program

The computer decision-making field is turning out to be somewhat similar to the production of automobiles. Engineers and designers work year-round to improve the product, but once a year they must freeze their progress and produce a model. That is the way it is in this field. Each new accomplishment opens the door to the development of another. We have reached a point in the commercial version where we intend to freeze and publish.

Before the actual availability of the program in a laboratory publication, we will make it available in either a listing or a card form to anyone interested in using it. Included will be instructions for its operation. Already several companies and equipment manufacturers have requested it, and we have furnished it to them.

DISCUSSION PERIOD

IRA L. LIBERMAN, vice president, Duke City Lumber Company, Albuquerque, New Mexico: Can systems be mixed—dimension and shop, for example?

HALLOCK: That is correct. You have the option of optimizing either 2-inch dimension only or a combination of 4-quarter jacket board and 8-quarter lumber from the balance of the log. And that jacket board, incidentally, may come both on the log and on the cant when you are cant sawing.

UNIDENTIFIED: Does the program take into consideration sweep?

HALLOCK: Currently, no.

UNIDENTIFIED: Have you worked with logs in the 30- to 40-inch size classes?

HALLOCK: No, we have not. This program is not designed to go much above 20 inches. Because of the breakdown methods we are using, we decided to work with the program on small logs only. However, the basic parts of the program—the sawing routines and the evaluation routines—could be adapted by someone who wanted to work with larger logs. A current limiting factor of diameter is that when you now get flitches wider than two 2x12s, the routine would not handle them and would require some changes.

UNIDENTIFIED: How is the program currently available, what source language is it in and also, how difficult would it be to modify it for use in a stud mill as opposed to the way it is working now?

HALLOCK: The program is written in Fortran 4. There are some things that are somewhat related to the machine. Anyone who is interested in the program, let me know and we will be glad to send you a source listing and/or a deck. With the source listing and/or deck, I will also send a list of places where I feel you may have to correct the program to adapt to a particular machine. If you are interested in using it on a particular machine, you might mention that, too. We know of five different machines it is running on now, so we might be able either to give you a listing for the machine you are interested in or to put you in contact with someone who could. It would not be very difficult to modify; we do plan more modifications. The 1974 model will be out before the end of 1973. It is our next production model.

5

Chipping headrigs & process requirements

James L. Gregoire, General Manager, Chip-N-Saw, Inc. Eugene, Oregon

Ernest Helvogt, Jr., Division Engineer, Crown Zellerbach Corp., Portland, Oregon

Part 1 by James L. Gregoire

The first Chip-N-Saw machine was installed about 1963. There are now more than 250 Chip-N-Saws operating throughout the world. We have machines in eastern and western Canada, on the U.S. West Coast, in the Rocky Mountains and in the South. We also have machines operating in Sweden, New Zealand and South Africa, and a machine has been operating for almost two years in the USSR. So we have had an opportunity to cut a wide variety of species, both softwood and hardwood, as well as species with which most of us are probably not very familiar.

Worldwide use does point out that the machine can process a wide variety of species. From the Mark I of 1963, we have progressed to the Mark II machine, the new Mark III machine, their canting systems and the canting systems used with twin band and quad band resaws. We have introduced the jumbo canter capable of taking logs up to 28 inches in diameter. We have progressed into a V-head core machine to take small-diameter cores as well as small-diameter logs. We have chipping edgers. All in all, we feel that we now have a well-rounded line of equipment to meet chipping headrig and chipping process requirements.

Outside kerf used

All of these machines are designed to eliminate the outside kerf normal in logs, thus increasing usable wood fiber. Let us go through some of the capabilities of the systems, starting with log orientation. In Wisconsin, I heard Dr. Dwight Hare, an ecologist with the U.S. Forest Service, Washington, D.C., enlighten wood-working machinery manufacturers on the economics of and the dramatic changes in our forest plans. He pointed out the difficulties sawmill people are having with ecologists who condemn the industry for not using all of the raw material available within the logs. He gave projections of the volume of logs and consumption in the year 2000, pointing out very vividly the need to find ways to improve recovery and get maximum yield.

Log orientation

Figure 5.1 shows two plans of log orientation. One configuration is called the full-taper; the other, the half-taper. We employed a consulting firm to make a computer study to determine which configuration would give the greater recovery in lumber from the log orientation. The study found that in logs from 5 to 11 inches in diameter, the half-taper configuration gave the greater yield. In logs from 11 to 20 inches in diameter, full-taper sawing gave the maximum recovery. Using this information and a scanner system, the log can be oriented to get maximum recovery by use of a computer.

Log diameters and taper factors should be programmed to allow the chipping heads to be set to obtain the best opening face for maximum recovery. Once this has been done with the log orienter, the scanner and the computer, let us look at some log profiles to see where we can gain additional usable fiber and to determine the recovery percentage in lumber, chips and sawdust.

Chips versus saw recovery

Figure 5.2 shows typical recovery on a 9-inch-diameter log 16 feet long with 1/16-inch taper per foot. We are going to use that particular size later in a study. From this log we are able to obtain a 1x4 14 feet long and six pieces of 2x4 16 feet long. The dimensions used here were 3/4-inch thick for the four-quarters, 1 3/4 inches on the eight-quarter and 3 3/4 inches in width. We have used these typical figures in lumber dimension totals to arrive at a total lumber volume of 68 2/3 board feet in which we were using a 1/8-inch kerf in the lines between the pieces. Here we can look at the amount of material turned into chips from the outside, which has zero kerf. The 1/8-inch kerf on the inside is lineally equivalent

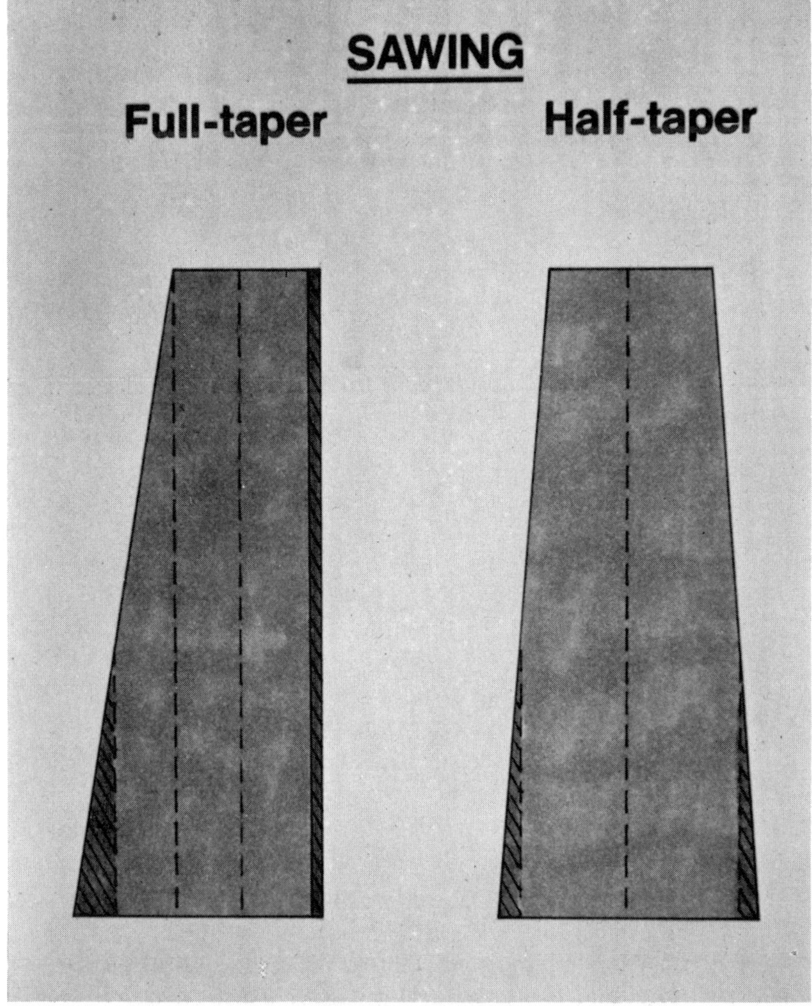

Figure 5.1. Two plans of log orientation: full-taper and half-taper.

to 22.5 inches, making total saw kerf out of this 9-inch-diameter log 22.5 inches. In lumber recovery we had 59.07 percent; 36.95 percent was in chips and only 3.98 percent was in sawdust. That is a typical low-sawdust production.

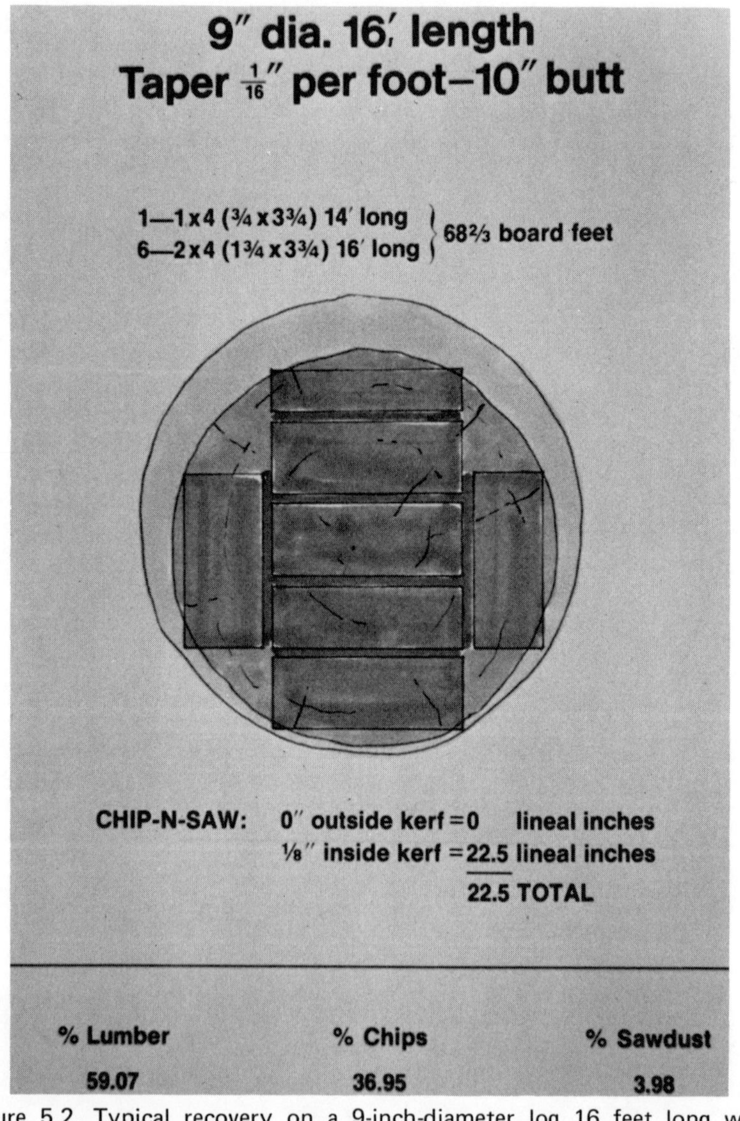

Figure 5.2. Typical recovery on a 9-inch-diameter log 16 feet long with 1/16-inch taper per foot.

Comparative yields

Figure 5.3 shows the identical log diameter, taper and board foot recovery as in Figure 5.2. We have sawn away the outside slab

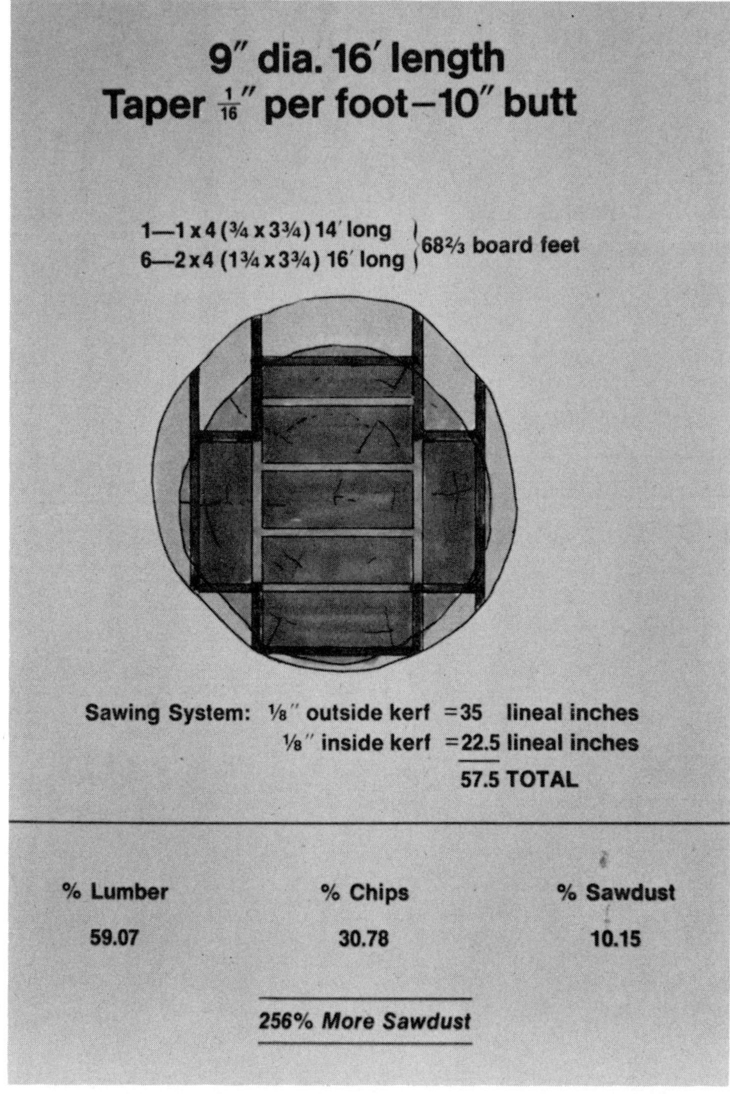

Figure 5.3. Sawdust figure is increased 256 percent as compared to the percentage of sawdust shown in Figure 5.2.

area as well as the edging. Again we are showing 1/8-inch inside kerf; in this example, however, outside kerf is 1/8 inch. Outside kerf amounted to a lineal equivalent of 35 inches, inside kerf again 22.5 inches, giving a total of 57.5 lineal inches of saw kerf on this 9-inch log. In lumber recovery we had again 59.07 percent; in

SAWING SYSTEM ON 9" DIA. LOG

Outside Kerf Variance

⅛" Kerf

% Lumber	% Chips	% Sawdust
59.07	30.78	10.15

3/16" Kerf

% Lumber	% Chips	% Sawdust
59.07	27.72	13.21

9/32" Kerf

% Lumber	% Chips	% Sawdust
59.07	23.08	17.85

CHIP-N-SAW

% Lumber	% Chips	% Sawdust
59.07	36.95	3.98

Figure 5.4. Production figures for three kerf widths are compared to production figures for a standard Chip-N-Saw installation.

chips, 30.78 percent and in sawdust, 10.15 percent—over 256 percent more sawdust in this log because we had to cut the outside slab area as well as the edging area.

Let us continue in Figure 5.4 with this same 9-inch log, but assume a different saw kerf. Assume 3/16-inch kerf both within and on the outer periphery of the log. (That is comparable to about a 14-gauge bandsaw kerf.) We again have 59.07 percent recovery in lumber, 27.72 percent in chips and 13.21 percent in sawdust—again a sizable increase in sawdust. On 9/32-inch kerf, which could be a very large band headrig with plenty of swage so the filer does not have to do much work, we had to go to a slightly larger log. But accumulating that into the same board footage of 59 percent, we end up with 23.08 percent in chips and 17.85 percent in sawdust.

Compare this yield with that of the chipping headrig where the wasted outside saw kerf line is eliminated. We have 36.95 percent in chips and only 3.98 percent in sawdust. If the industry is to find better yields of usable wood fiber, it should be apparent that these new chipping headrig systems offer this increased utilization.

Chipping edger

Figure 5.5 takes a look at the chipping edger in respect to chipping versus sawing. Here you see two pieces of 2x8 16 feet long in which we have calculated the amount of sawdust on a variable saw kerf in pieces per minute, as a basis for comparison. We have talked in the past of accumulating that in total lineal inches. Here I have tried to put it into something more realistic to lumbermen—in piece count per minute to edgers. I have prepared the figures based on 10 pieces per minute, 15 pieces per minute, 20 pieces per minute and 25 pieces per minute and calculated the volume of chips per day that the kerf would amount to based on 400 minutes and 220 working days per year. The figures show the amount of usable wood fiber available with these systems.

Maximizing recovery

Let me assure you that we in the machinery business are concerned with the lumbermen's raw material supply. We are directly affected by ability of lumbermen to make a profit. We have acquired a new plant in Eugene, Oregon, and are banking on lumber industry success.

Let us take a mill summary that more dramatically points out usable wood fiber. Figure 5.6 takes a sawmill that produces 40

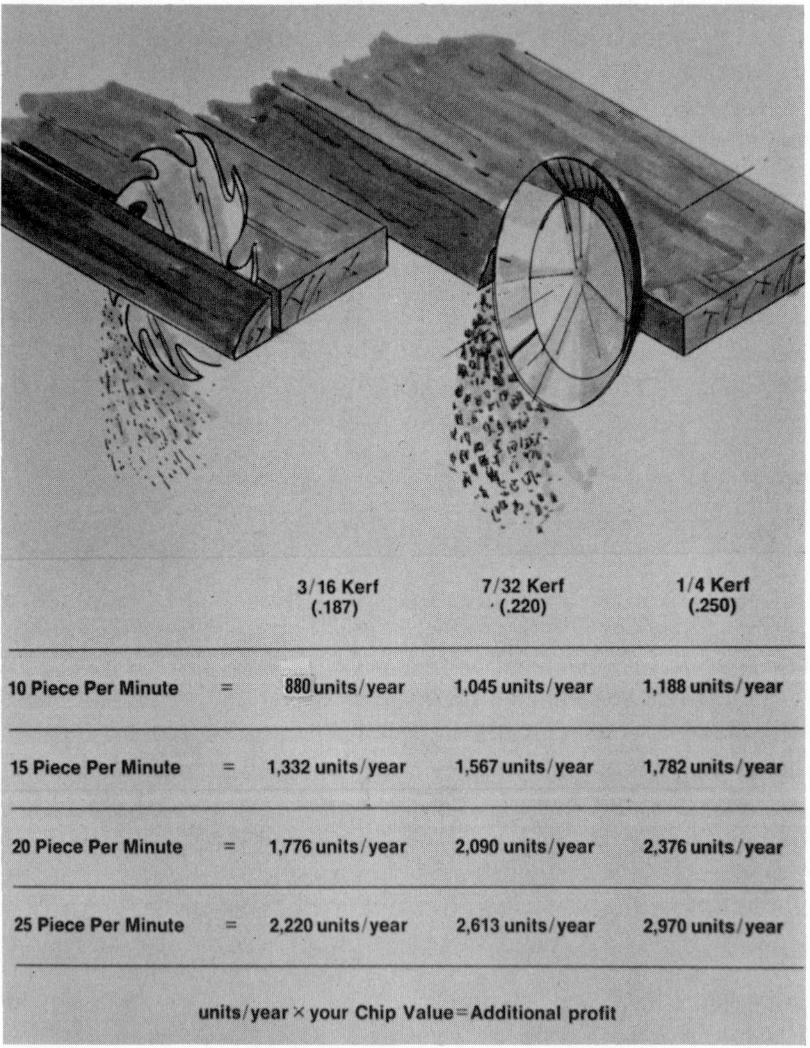

Figure 5.5. Using three different saw kerfs, these figures show the total volume of chips lost to sawdust in one year at the edger.

million board feet per year on an average weighted diameter again of 9 inches. Base calculations on usable fiber recovery. Go back to the original profile design in which we had the very low sawdust production of only 3 percent. What does the usable fiber obtainable through a chipping headrig amount to in chip recovery on 40

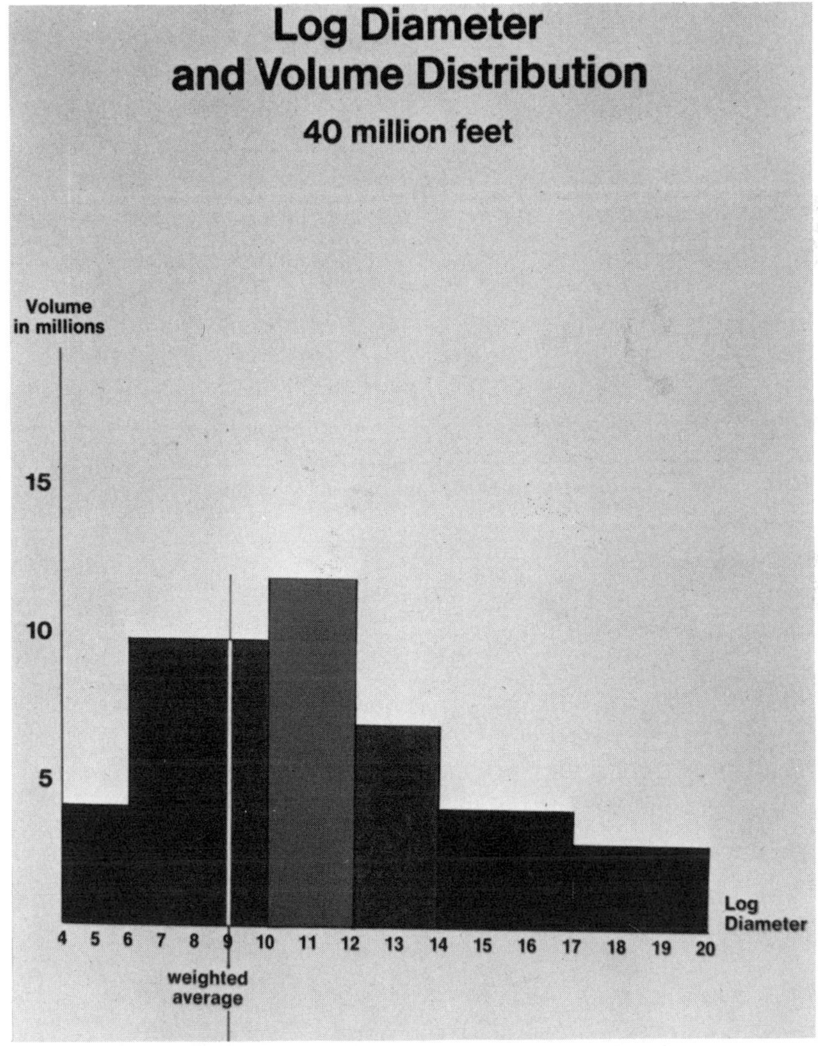

Figure 5.6. Log diameter and volume distribution for a sawmill that produces 40 million board feet a year on an average weighted diameter of 9 inches.

million board feet—based on a return on the chips of from $15 to $25 per unit? Figure 5.7 estimates the additional chip recovery and income. On the sawing system we recovered 19,413 units; under the chipping headrig system, 23,442 units—a gain of 4,029 units per year. At $15 to $25 per unit, the bottom line figure gain is dramatic. Chipping headrigs and chipping edgers eliminate wasted sawdust around the outside periphery of the logs and can

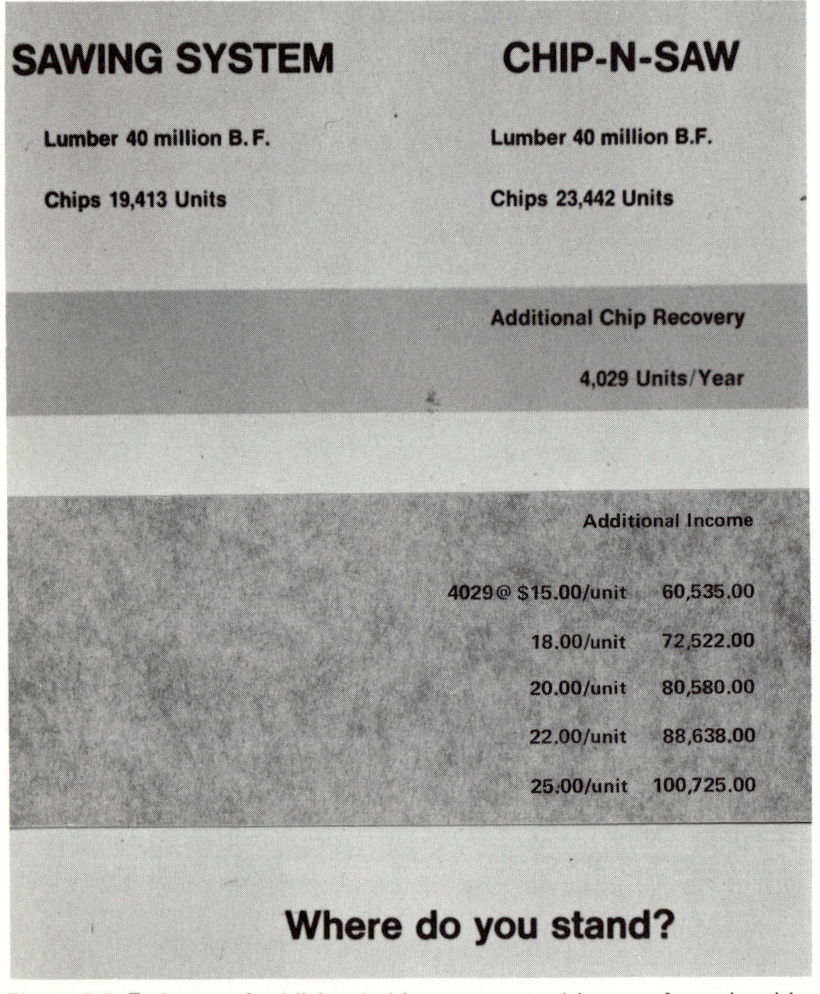

Figure 5.7. Estimate of additional chip recovery and income from the chipping headrig system as compared with the sawing system.

return profits over the sawing system by use of scanners and computers so that logs can be oriented and machines programmed to obtain the best opening face. This is getting down to the bottom line by maximizing yield with modern machines and concepts. We feel at Chip-N-Saw that we are on a new plateau of efficient machines that maximize recovery.

Part 2 by Ernest Helvogt, Jr.

At Crown Zellerbach's operation in Columbia City, Oregon, the Chip-N-Saw was installed September 1, 1968. It is a Mark II quad with constant speed of 120 feet per minute. Operating capacity is 5-inch minimum diameter to 16-inch maximum diameter. The present operating range is from a 6-inch minimum to about a 14-inch maximum. Species breakdown is about 95 percent coastal hemlock and about 5 percent fir. When the machine was put in, the breakdown was about 60 percent hemlock and 40 percent fir. The mix has changed. Average mill log diameter is 10½ inches, average log length about 18 feet. Average machine diameter on the Chip-N-Saw is very close to 9.4 inches and no more.

The Columbia City mill is a two-line, small-log sawmill, covering a diameter range from 6 to 26 inches, with 95 percent under 19 inches. One side consists of a Chip-N-Saw quad, an 8-inch Schurman double arbor edger, a board edger and a Prescott trimmer. This side cuts the range from 6-inch minimum to 14-inch maximum. The second side consists of a Mill Engineering log gang, cant gang, board edger and a trimmer. It cuts a range from 14 to about 26 inches. The two lines combine at the outfeed of the mill and go into a 14-tray sorting system. Total mill output is 190,000 to 200,000 board feet per shift.

Diameter critical

The reason for the range in the output is that the product varies directly with log diameter and/or length. In this mill, which runs specifically on a piece count, if diameter range varies about one inch, you can expect somewhere in the range of a 30 percent decrease in production. Diameter is critical.

Low maintenance and high recovery

This Chip-N-Saw quad has run two shifts per day since installation, with minimum repair and upkeep. As nearly as we can tell without detailed costs, which we do not have, maintenance appears much less than combined maintenance on the two gang saws that run in the adjacent line.

Mill recovery is measured on board foot lumber per cunit of logs. With this mix of logs, recovery can vary from 600 to 800 board feet lumber per cunit of logs. The mill now cuts 2x4, 2x6, 2x8, 2x10, 1x4, 1x6, and some 2x3 as a falldown item. With this mix, the mill yields an average of about 750 board feet lumber per cunit of logs. Now if you were also to cut 2x2 and 2x3 within your cutting pattern, I believe that recovery would increase to an average of about 830 board feet or to, say, a range somewhere between 800 and 900 board feet, depending on the diameter range.

To achieve this type of recovery, you must determine cutting patterns for this type of machinery based on the following factors—green cutting sizes for 1-inch should be about 12/16 inch; for 2-inch, cut about 1 11/16 inches.

Cutting pattern is determined from the dry American Lumber Standards (ALS). Full wane allowance on finished lumber sizes is projected into the green cutting pattern. Log taper is allowed to run out the wane. Recovery is usually determined by one of two methods. The first is cubic volume, which is based upon two constants—board feet of lumber and cubic volume of logs. This is the more accurate method to compare cutting methods or differences between mills.

The second method employed to measure recovery is overrun, which has been used for many years based upon Scribner scale. It does have some disadvantages. There are many variables:

1. Type of mill—cutting mill, dimension mill, stud mill.
2. Log scale—long or short.
3. Type of scale—Columbia River, International, Forest Service, et cetera.

It is not uncommon for someone to ask "What's your overrun?" But one mill that has 50 percent overrun on a dry basis is really quite comparable to a mill that has 100 percent overrun on a green basis. There is that much disparity in the range.

Scanner system installed

The mill at Columbia City is installing a scanner system, which should increase recovery substantially. We will scan both lines of logs—those coming for the Chip-N-Saw side and those for the gang side. Based upon the scan information and through the mini-computer, we will study the Chip-N-Saw. We will also be interfacing mill lumber output from the existing memory system into the computer. We will then have a printout—a shift-end report for mill operating control—for the man on the floor and for statistical information for accounting.

The advantages of this small-log breakdown unit are:

1. Linear-type flow results in much higher throughput than conventional headrig and carriage, which have a reciprocating action.
2. Chipping the slab on initial breakdown eliminates further handling of material that is really not used as lumber.
3. Plug-ups and the clean-up problem of pieces dropping on the floor are eliminated.
4. A higher degree of control on log breakdown results in straighter lumber and a straighter cant.

As a result of chipping and cutting equally and progressively from each side of the log, lumber and cant seem to be relieved of stress. This results in a straighter piece. There is less twist and sweep, and propellering is practically eliminated in the finished product. When using a quad or a twin behind the chipping section to cut boards, thin saws can be used to a great extent. At Columbia City, our saw kerf is 0.125; 1/8-inch kerf has been working very well.

6

Double-taper chipper canters

Gilbert W. Anderson, President, Stetson-Ross
Machine Co., Inc., Seattle, Washington

John L. Ailport, Plant Manager, Diamond International
Corp., Newport, Washington

Part 1 by Gilbert W. Anderson

The attendance at this Sawmill Clinic certainly attests to the industry's interest in obtaining the greatest recovery possible from our forest resources. We feel that the double-taper chipper canter, the small-log machine to be described, provides a unique opportunity for industry to obtain this greater recovery.

What do we mean by the term *double-taper* approach to chipper canters? It is very simple. We are providing a machine that capitalizes on natural taper in the log to produce more lumber than is obtainable from other types of cutting.

The impetus to design such a machine came in discussions with representatives of Diamond International Corp. in the latter part of 1971. Diamond International was looking for a way to obtain greater lumber recovery than was possible with existing machinery; they wanted to incorporate such a method into designs for a mill to be built at Albeni Falls, Idaho. With experience in high-speed chipperhead applications to small logs, Stetson-Ross felt it could answer Diamond's challenge.

A short film describing the machine is available to the industry. The material that follows this brief introduction is actually taken from the film narration. The reader of this paper will not have the

benefit of the film but, of course, the accompanying illustrations
(Figures 6.1 through 6.5) are drawn from the film and will serve to
illustrate the main points involved in the operation of the equipment described in the narration.

Anderson's narration of the film

Stetson-Ross has been manufacturing machinery for the lumber
and woodworking industries since 1898. One of our biggest installations was recently completed for Diamond International's new
Albeni Falls facility. Diamond had decided to build a maximum-
yield sawmill and planing mill complex that would derive most of
its production from logs less than 20 inches in diameter. With saw
technology that is able to hold closer and closer tolerances with
very thin kerfs, the natural taper in these small logs becomes a
source of additional lumber recovery. To realize this recovery,
however, logs of this size must be run very efficiently at feed rates
substantially greater than those typical of the conventional sawmill operation.

Diamond personnel set about designing a mill that would produce as much as 150,000 board feet a shift from logs averaging less
than 15 inches in diameter. To do this, they selected the Stetson-
Ross chipper canter as the primary breakdown machine.

A high-speed log infeed system receives logs from the yard.
They are processed through a debarker, then cut to length. Logs
enter the mill on a log haul that conveys them to the Stetson-Ross
double-taper chipper canter. The log is positioned in the four chipperheads, which cant the log by chipping four parallel faces on it.

Cant to twin resaw

The cant is fed immediately into a fixed linebar twin resaw for
further breakdown. The theory of double-taper canting is to orient
the log as on a conventional sawmill carriage. In the Stetson-Ross
machine, a fixed bottom and fixed side head chip full-length faces
on the log. These opening cuts are oriented 90 degrees from each
other (as indicated by the heavy dotted lines in the lower right-
hand corner of Figure 6.1). Taper can then accumulate across the
horizontal axis from the fixed side head as well as along the verti-

cal axis from the fixed bottom head of the machine.

A top chipperhead is set to chip a face parallel to that opened by the bottom head, and the other side head is set to open a face parallel to that chipped by the fixed side head. If the log has significant taper, these selectively set faces will be less than full length (as shown in Figure 6.1). The result is that additional lumber can be recovered from the taper that is then accumulated along two axes.

Few shorts made

Lumber recovered will include a minimum of shorts as the taper is oriented to affect only two boards. With two full-length faces

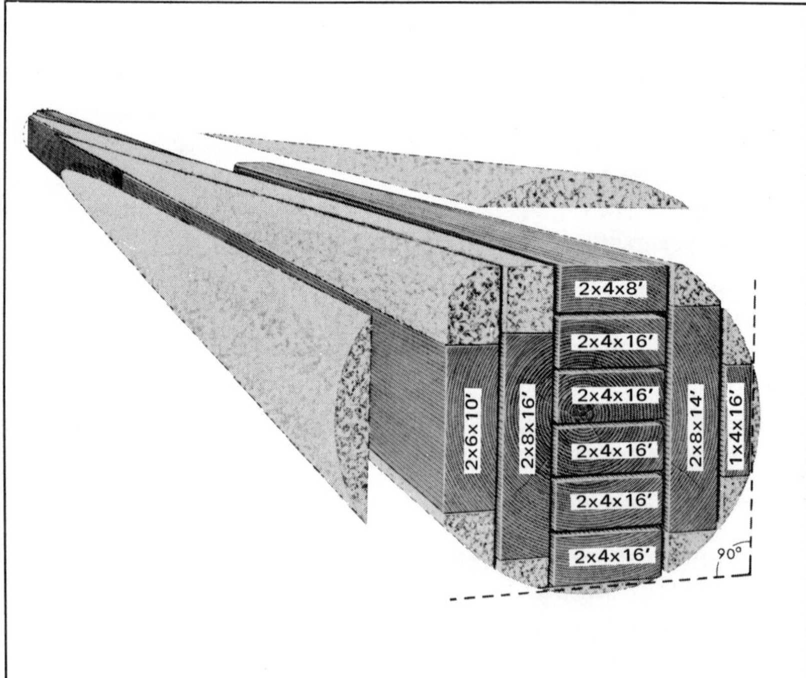

Figure 6.1. The diagram shows the yield from a 16-foot log of 10¼-inch top and 12¼-inch butt diameter, based on two full length opening faces, 90 degrees apart. Cutting diagrams showing other yield patterns for various log dimensions are available.

oriented 90 degrees from each other, the cant can be accurately broken down into lumber because the bed and guide reference has been established the entire length of the log.

Log-orienting device

The canter is preceded by a log-orienting device consisting of three rolls that can be skewed to position a log to best advantage and are powered to feed the log into a buckhorn chain. The flights of this chain are arranged to provide a vertical and a horizontal index against which logs are firmly held by pressured rolls. The entire feedworks can be raised and lowered along a 45-degree axis to expose more or less of the fixed chipperhead. This is done to adjust for the appropriate depths of cut for different sizes of logs to establish the best opening face.

Four climb-cutting chipperheads are embodied in the chipping section and arranged with the bottom head and right side head fixed and the top head and the other side head movable with setworks. Powered outfeed rolls pull the cant through the chipperheads and send it downstream for further breakdown. The full-length face chipped by the bottom head continues on a powered roll chain and the face chipped by the right side head is forced against the fixed linebar by powered crowder rolls and fed on through a twin resaw.

One-man operation

This entire system of machinery is operated by one man. Each log is programmed for one of 15 cutting schedules. One button on the operator's console causes the canter infeed, two chipperheads and the twin bandmill to set the positions that have been programmed on matrix boards. The programming is easily changed by repositioning diodes.

Cants from the twin resaw requiring further breakdown are transferred to a quad resaw that will also receive flitches from the headrig side of the mill. Lumber requiring edging will be routed to one of two Stetson-Ross chipping edgers. These machines, like the canter, utilize circular-bit chipperheads designed with a cutting angle and radial gullet for optimum chip quantity and quality.

Chips highly regarded

Chips from these heads are considered excellent for most pulping operations. The chipping edgers incorporate splitter saws for upgrading wider material.

Green lumber from the mill is trimmed and passed through an indexing bin drop sorter. From there it is stacked, dried and routed to the planing mill. Most sawmill production will run through one 20-knife Stetson-Ross planer matcher capable of 1,200 lineal feet per minute. Finished lumber is then ready for shipment.

Logs are becoming smaller and smaller as modern forestry techniques continue to accelerate the timber harvest from sustained-yield forests. This complex is designed to take maximum advantage of our most renewable raw material.

Curved knives used in the Stetson-Ross canter have clean cutting action. In Stetson-Ross's cutter edge research and development program, an effective way to determine how cutterheads are performing is to take high-speed pictures of them. For example, when a camera is running at about 10,000 frames per second, it is possible to monitor knives cutting at 1,800 revolutions per minute with boards advancing at 680 feet per minute. We ran tests on frozen boards with equally good results—chips were good and edges quite smooth. An inherent advantage of the Stetson-Ross line of canters is the positive chip control from the shape of these knives. I think the amount of chip overspray in a Stetson-Ross canter is a minor waste product.

In addition to the double-taper canter, Stetson-Ross supplied Diamond with a 24-inch chipping edger and a vertical arbor gang saw plus all conveying equipment. Bandsaws were manufactured by Letson & Burpee, Vancouver, British Columbia, and were supplied as part of Stetson-Ross's basic contract with Diamond to insure proper interface between the canter, the bandsaws and the edgers. The indexing bin drop sorter was supplied by Lumber Systems, Inc., Portland, Oregon.

Test run time

When we completed the first double-taper canter, we set it up in our plant and ran it for 30 days. I think this is the first time that we, as a machinery manufacturer, ever truly appreciated start-up and electrical costs. Local timber companies provided logs for the test runs in Seattle. We were able to establish the canter's capability to handle mill-run logs since our local friends had been very careful to supply us with low-grade logs. We had a good chance to process bananas and swelled butts and schoolmarms of all descriptions. We did another interesting thing in bringing hardwood logs from the South, so we were able to process successfully gum, oak, cypress and hickory. Surface on the hardwood cants was extremely good, with minimum tear-out.

We produced the machine in two models to cover a range of log diameters from 3½ to 24 inches. Minimum cant size is 3 by 4 inches; maximum is 20 by 21 inches. The machine can operate from 93 to 325 feet per minute, processing 4,000 logs per shift with relative ease. The chip length is 5/8 inch to 1 inch.

Results analyzed

We had a computer analysis of the machine performed by an independent engineering consulting firm. The purpose was to compare yield results of the double-taper machine with the other four-head chipping headrigs working on the same size-range of logs. This analysis indicates that in logs over 7½ inches in diameter, the double-taper approach will provide a substantial increase in yield over these other chipping headrigs.

An 8 3/4-inch-diameter log will produce 82 board feet using the double-taper approach (Figures 6.2 and 6.3). It is possible to produce 68 2/3 board feet from a 9-inch-diameter log using the other basic approach. There is a considerable difference between these two methods of chipping.

We have copies of the computer study and would be more than happy to review it with industry. This is truly a new concept in chipper canters. Diamond International was the pioneer in accepting the concept and has the first machine. Diamond has been joined, however, by Champion International Corp., which will install a double-taper chipper in its Bonner, Montana mill, and by Kimberly-Clark Corp., which ordered a unit for its Waynesboro, Georgia mill. At Kimberly-Clark, 40 percent of the production through this machine will be hardwood; the balance of the production will be southern pine.

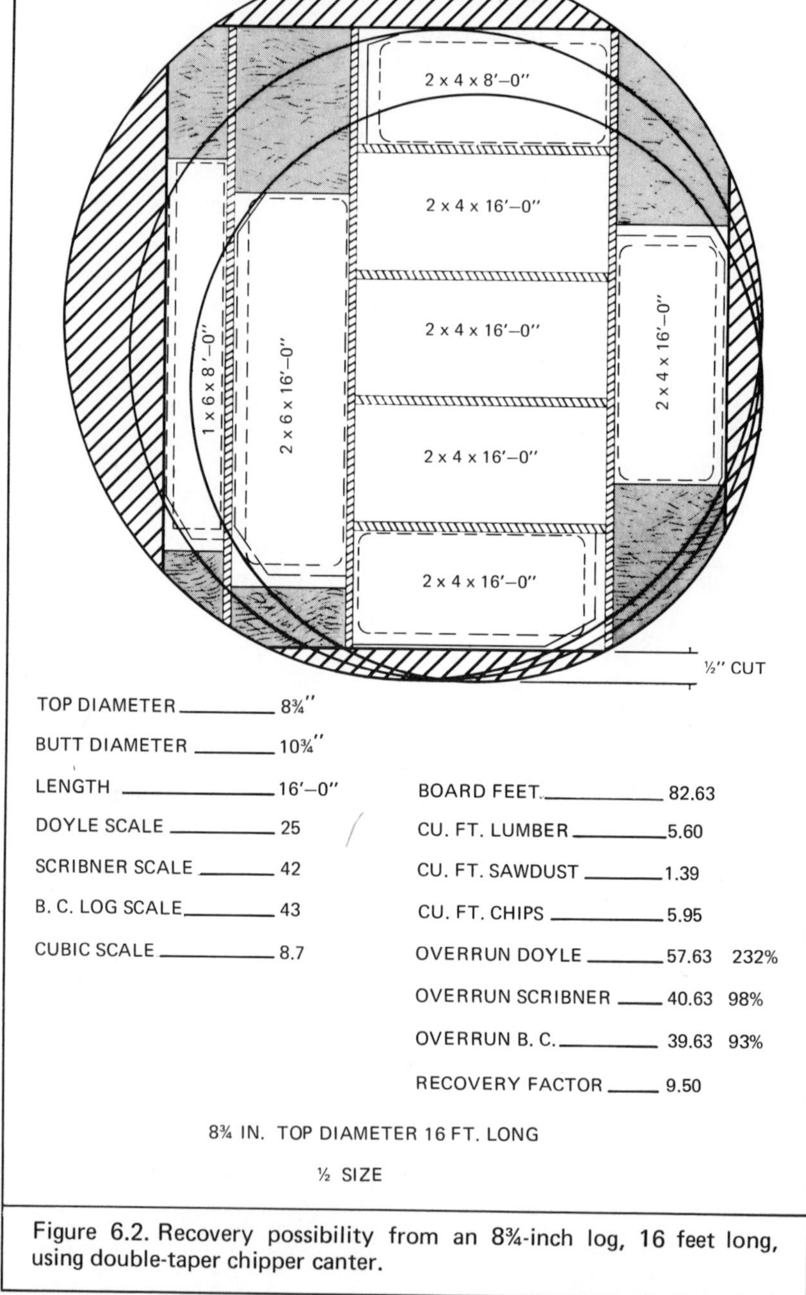

Figure 6.2. Recovery possibility from an 8¾-inch log, 16 feet long, using double-taper chipper canter.

CHIPPER CANTER YIELD COMPARISON
CUTTING SCHEDULE ASSUMPTIONS
DOUBLE-TAPER CHIPPER CANTER

Small circle represents top of log.
Large circle represents butt of log.
Dotted circle represents middle of log.
Taper is 1/8" per foot or 2" in 16'.
Kerf 1/8" – Bandsaw Lines 1/8" – Circular Saw Lines

Nominal	Green	Shrinkage	Dry Surfaced
1 x 4	15/16 x 3 7/8	1/32 x 5/32	3/4 x 3 1/2
1 x 6	15/16 x 5 7/8	1/32 x 3/16	3/4 x 5 1/2
2 x 3	1 3/4 x 2 3/4	1/16 x 3/32	1 1/2 x 2 1/2
2 x 4	1 3/4 x 3 7/8	1/16 x 5/32	1 1/2 x 3 1/2
2 x 6	1 3/4 x 5 7/8	1/16 x 3/16	1 1/2 x 5 1/2
2 x 8	1 3/4 x 8	1/16 x 7/32	1 1/2 x 7 1/4
2 x 10	1 3/4 x 10	1/16 x 1/4	1 1/2 x 9 1/4
2 x 12	1 3/4 x 12	1/16 x 9/32	1 1/2 x 11 1/4

Length each board noted.
Allowable wane used is 1/3 width of any face and 1/4 the length accumulative.
Finished planer cuts – 1/32 bottom, 1/16 right side, 1 1/8" eased edge radius.
Lumber shown is standard or better per P.L.I.B. grading rules for wane.
1 cu. ft. solid wood = 2.5 cu. ft. sawdust.
1 cu. ft. solid wood = 2.78 cu. ft. chips.
Overrun shown = board feet lumber vs. log scale.
Recovery for dimension lumber except 1" boards as they develop.
Recovery factor shown is nominal board feet of lumber divided by cylindrical volume in cubic feet defined by the median log diameter.
Shortest significant board – 8'.
Thinnest significant board – 1".
Narrowest significant board – 4".

Figure 6.3. Cutting assumptions for the yield shown in Figure 6.2.

Part 2 by John L. Ailport

Before we at Diamond International had tapers on rigs, we started with the log on a set of blocks and knees and cut from one side. So essentially we are coming back to exactly what we used to do when we speak of the double taper. We cut the one side, then turned the log over so the other side was parallel to the saw. Then we squared the cant in the same manner and again we were parallel to the saw on the one side.

Wood wasted

This was not a good system in those days because the logs were

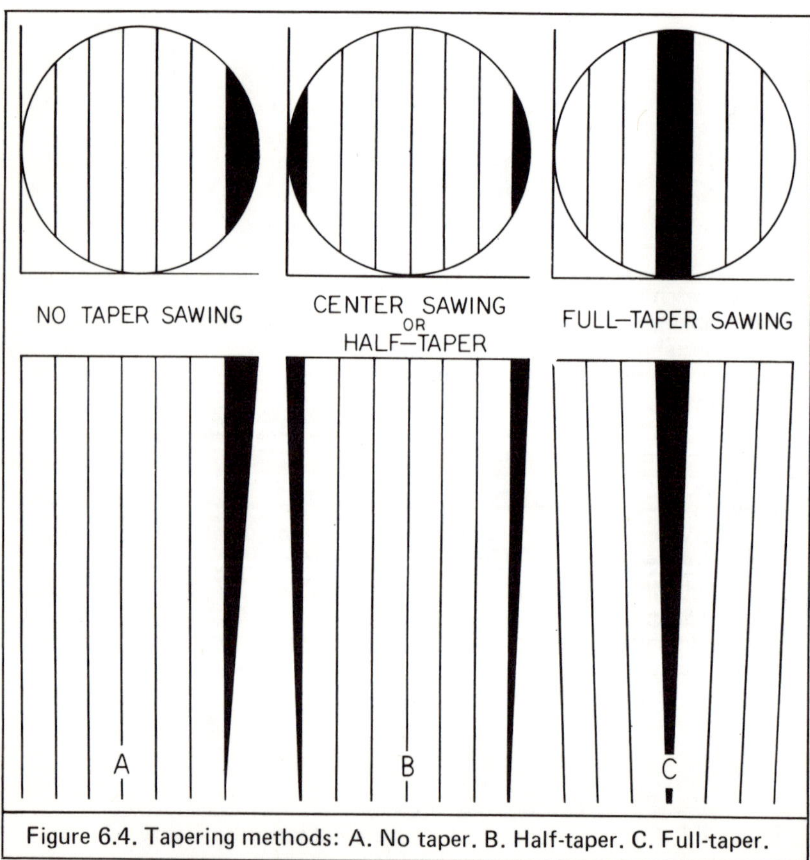

Figure 6.4. Tapering methods: A. No taper. B. Half-taper. C. Full-taper.

larger. Many sawmillers my age can remember when 6-inch tops
were just limbs. Later we got tapers on the rigs and got into
tapering for better yield and for better grade. Now we can center-
saw or half-taper (Figure 6.4). We might speak of center-sawing
when talking about canters. In the early days when we got tapers
on the rig, we would talk about half-tapering or full-tapering. Most
people went to half-tapering for two reasons: (1) time was limited
and (2) raw material was not as important. So we would cut one
side half-tapered, pull the taper out and the other half would be
automatically taken care of. This was fine when we did not have
to worry about maximum utilization. We got longer boards by this
method.

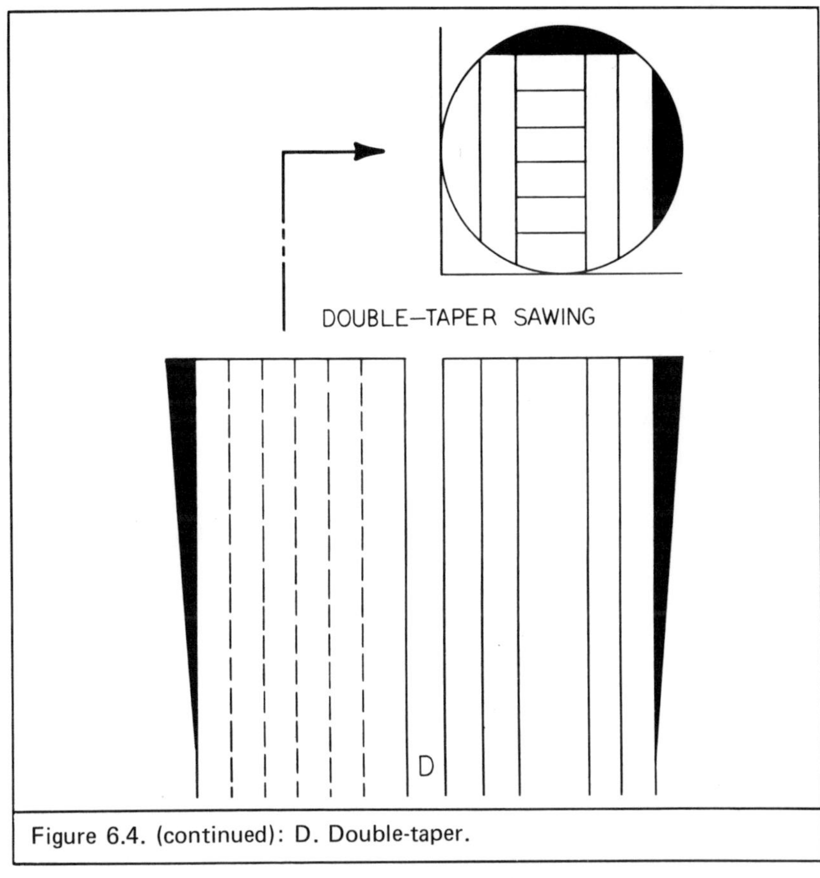

Figure 6.4. (continued): D. Double-taper.

Recovery raised

Then we started talking about more grade from the log and we got into full-tapering. Some mills never have full-tapered; others have. There are two good reasons for it, providing the logs are the proper size, and there are several different methods of taking the taper out. If we cut parallel to the outside of the log, we get longer selects. We get longer shop boards when cutting ponderosa pine or timber of that type. We could take the taper out in back of the shop boards in the lower grade material or, if the hearts are bad, we can take the taper out in the center. This is still an excellent practice on big logs. We get wide common boards out of those center cuts rather than short selects on the side. We also get into situations where you have selects on one end of the board and

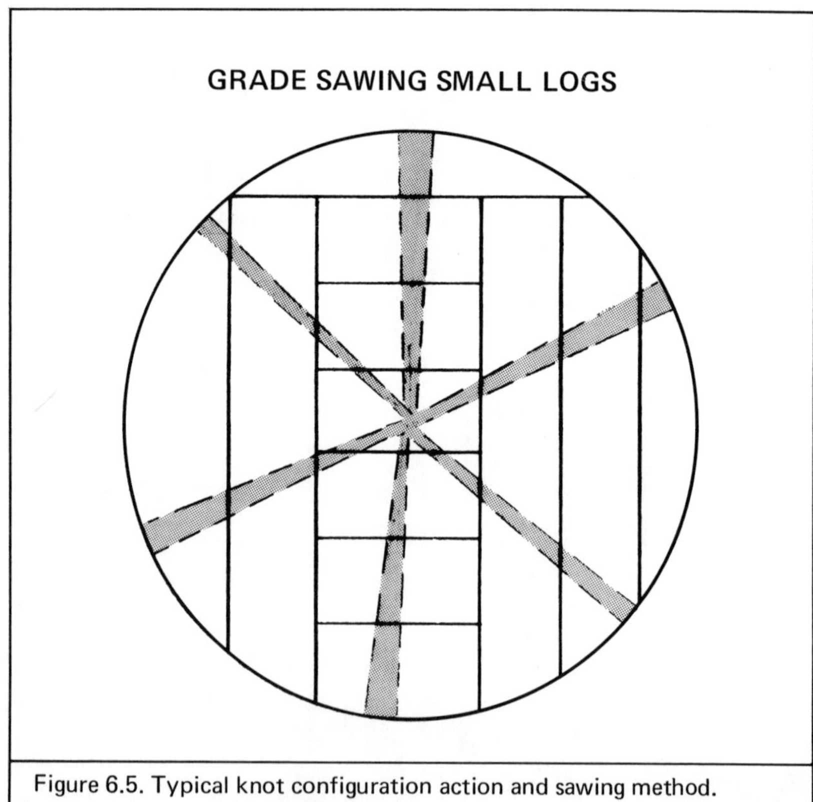

Figure 6.5. Typical knot configuration action and sawing method.

commons on the other where you are not tapering. We wind up with both short commons and short selects in this case.

Two faces opened

The term *double-tapering* is to me a misnomer because it is actually no taper at all. It is a name. We are doing exactly what we did back in the days when we did not have tapers on the rig, but in this case we are doing it in one pass. We are able to cut both faces at the same time. We are parallel to the bottom and the side of the log and we are able to take advantage of the 1½-inch taper which is normal in some geographic areas. I think this will vary in different parts of the country. I can think of areas in Montana where you would have 3 or 4 inches of taper. On the West Coast, there might be less. But where we are in north Idaho, 1½ inches in a 16-foot log is the average taper. By using this method we gain boards on the tapered side. We are developing short 6-, 8-, 10- and possibly 12-foot boards on the side opposite the so-called double taper.

In our mill at Albeni Falls, we use a scanner to determine log size. On the various sets, this will tell us which size setting will give us maximum utilization. In some instances, we are using a center cut. By center-cutting, we can improve the grade. All lumbermen know how knots are oriented in a log (Figure 6.5). Taking the center portion of the log—which can be 4 or 6 inches (I prefer 4 inches)—we are cutting across the knots except for the center board, or maybe two boards, where there would be branch knots or spike knots. If we were to cut the board the other way, we would have long branch knots and a much weaker piece. Figure 6.1 shows the combination of a better grade and the additional footage that you pick up on the side of a log. Better grade is picked up in the outer boards and the additional board is picked up on the top and the side in shorts. So we are getting more dollars out of the grade, more dollars out of the footage.

Production in this type of sawmill is increased because we are running logs straight through the machine. There is no backing up, no reciprocating action—we just keep it going straight ahead. There are logs going in one end and boards coming out the other. Accuracy in making a cant is another advantage of the so-called double-taper machine because the cant is held against a solid bedplate and a solid guide. This is the method that we have used in planers for many years. Use of multiple bands allows us to slow down the speed so we are able to improve the sawing accuracy. Accuracy, combined with thinner kerf, gives greater yield.

To solve problems by complicated methods and exotic equipment takes very little real effort, but to develop a simple method and uncomplicated equipment takes real thought. As we look at the number of people searching for new ideas and at the strides we have made in the last few years, it gives me great hope that lumbermen are really on the move—really ready for change.

I would like to conclude with a poem that I have had for many years and have always thought appropriate.

The Primeval Calf Path

One day through the primeval wood,
A calf walked home as good calves should.
And made a trail all bent askew,
A crooked trail as all calves do.
Since then three hundred years have fled
And, I infer, the calf is dead.
But still he left behind his trail,
And thereby hangs my moral tale.
The trail was taken up next day
By a lone dog that passed that way.
And then a wise bellwether sheep
Pursued the trail o'er vale and steep.
He drew the flock behind him, too,
As good bellwethers always do.
So from that day, o'er hill and glade,
Through those old woods a path was made.
And many men went in and out
And dodged and turned and bent about.
They uttered words of righteous wrath
Because 'twas such a crooked path.
But still they followed—do not laugh—
The first migrations of that calf.
Each year, a hundred thousand men

Follow this zigzag calf again.
And on its crooked journey went
The traffic of a continent.
A hundred thousand men were led
By one calf who thrice a hundred years was dead.
They followed still his crooked way
And lost a hundred years a day,
For thus such reverence is lent
To well-established precedent.
A moral lesson this might teach
Were I ordained and called to preach,
For men are prone to go it blind
Along the calf path of the mind.
They work away from sun to sun
To do what other men have done.
They follow in the beaten track
And in and out and forth and back;
And still their devious course pursue
To keep the path that others do.
They keep the path a sacred groove
Along which all their lives they move.
And how the old wood gods laugh
Who first saw that primeval calf.
Ah, many things this tale might teach,
But I am not ordained to preach.
Anon

DISCUSSION PERIOD

BOB VADNAIS, Totem Equipment Co., Seattle, Washington:
Does it make any difference in your operation whether a log is
turned? In other words, if you had the small end going
through all the time, does that make any appreciable difference in the operation?

AILPORT: No, it does not matter. In our scanner setup, the scanner will read the log diameter 8 feet from the butt end, regardless of which end is first.

SKIP LUHR, Wasatch Forest Products, Inc., Evanston, Wyoming:
Those logs were awfully straight and smooth. What about
sweep and knots?

AILPORT: Those logs were normal logs. We have good ones!
Sweep? We have sweep as well as anyone else. We are, however, cutting out as much sweep as possible. In the double-
canter machine, we will not take logs with a very heavy sweep.
We have a conventional headrig in the mill so we will move
that log to the other side. In the case of knots—even large
knots—our method of rolling the log orients it in such a way as
to take care of a knot that might push the log off center.

**WALTER COTTEN, Packwood Lumber Co., Packwood, Washing-
ton:** What kind of overrun were you able to establish with this
type of machine and what scale would we be talking about?

AILPORT: Well, we just started this mill, so I am reluctant to
mention overrun.

7

Advantages of headrig chippers

Fred Miller, Chief Engineer, Chipper Machines &
Engineering Corp., Lake Oswego, Oregon

Frank Roppel, Plant Manager, Ketchikan Spruce Mills,
Inc., Ketchikan, Alaska

Part 1 by Fred Miller

Since Chipper Machines & Engineering Corp. first introduced the headrig chipping principle based on the Standal patents, the headrig chipper has become well known in the sawmill industry. It is becoming almost as important as the headrig itself. This is quite obvious when mills using a single-cut headrig consistently report 15 to 30 percent production increases from equipment that costs as little as $60,000 to $80,000 to install.

The advantage of a double over a single cut is reduced substantially when a headrig chipper is added to a single-cut rig. The double cut, when supplemented with a headrig chipper, will generally show about a 15 percent production increase. Because of the additional double-cut requirements, cost of the headrig chipper will be slightly higher.

Importance of the headrig chipper was brought out a few years ago in a Canadian mill that had broken a band wheel. While the wheel was being replaced, the mill continued operating with the slabber. By putting the slabber up to the zero line, they were able to maintain approximately 60 percent mill production without a headrig, which helped them reduce the time lost, at least to a degree. Logs were slabbed and then resawn or sent to a gang.

Advantages

There are several tangible advantages—in addition to the increases in production—that occur with this type of equipment. Obvious advantages are conversion of saw kerf to chips and elimination, in many cases, of the need for a tail sawyer. Intangible gains are many. Slabs coming from flares, burls and crook are removed before they can jam conveyors and slab dropouts.

This material can be converted into quality chips whereas previously it was hogged, burned or converted into chips of questionable quality. We hesitate to discuss the possible increase in chip production resulting from the actions of an overenthusiastic sawyer. This does happen occasionally, particularly during indoctrination. Sometimes it is fun just to watch the chips come out.

We are certain, however, that it is only occasionally that a board gets into the chip bin. One mill reported chip production up about 12 percent, which seems to be a fairly reasonable figure. Saw life is

In its Dwyer Division sawmill Publishers Paper installed a 250-horsepower CM & E Model 64 headrig chipper. Working in conjunction with a 9-foot bandmill, the chipper has reduced the piece count 5 to 8 percent while production has increased 8 to 10 percent. The board coming off this cut is from the first saw line. Tail sawyer was retained to offbear and route material.

Photo courtesy FOREST INDUSTRIES magazine

improved because the in-and-out action of the saw in a thin slab is reduced. The wear effect of a saw going in and out of a thin slab is progressively aggravated. It gets worse as the action continues, due to the unequal wear on the two sides of the saw. With the slabber, it is possible to maintain fairly uniform thickness on the first board.

Improvement of chip quality

Contrary to the belief that variations in carriage speed will be detrimental to chip quality, we find in most cases that slabber chips will upgrade the general quality of the chips from a mill. In spite of moderate changes in feed rate, slabber chips will normally fit pretty well into a standard chip specification. Be sure, though, that the sawyer is aware of the need to maintain a constant speed. Many sawyers play with the controls, unless the need for a constant speed is definitely pointed out to them.

One important feature concerning chip quality from a headrig chipper is the stability of the slab by virtue of its attachment to the log. It is difficult to get this type of support in a wastewood chipper.

New outlook for sawyer

Another advantage is that the sawyer can devote more time to sawing. In most cases, he is concerned about the offbearer, the tail sawyer. He must take care that the offbearer is not in a dangerous position.

We are often asked about relocating the sawyer to a position over the slabber and about the sawyer's reaction to the added controls necessary with a headrig chipper. We found that, after initial indoctrination, most sawyers prefer this position. They have a better opportunity to see what is happening. They are not in the immediate path of a log if it gets knocked off the carriage. In general, I believe they enjoy the better view from this position.

Controls combined

To reduce the sawyer's work, we have been able to combine

controls and automate to the extent that, even with a double cut, the additional controls may be incorporated in a simple selector switch. We have had many favorable reports in regard to the speed with which the sawyer is able to retrieve previous production levels. This has happened in as little as four hours. At that point, the sawyer's further familiarization with the equipment leads only to a production increase.

Other elements of the installation advantageous to the sawyer are automatic setting of the saw guides and the installation of some carefully positioned lights to produce shadow lines or light lines. An air-conditioned and soundproofed cab, fitted with television, may contribute more toward increasing production than many of the other things mentioned.

At a Southwest Forest Industries sawmill a CM & E headrig chipper was installed to increase utilization from small-diameter pine logs. The goal in this operation is to send 40 percent of the mill volume to twin 5-foot bandmills as cants. The fully enclosed sawyer's box is located above the chipper and the loading deck, providing a better view for the sawyer.

Photo courtesy FOREST INDUSTRIES magazine

Handles eight-foot log

Headrig chippers are available in a broad range of sizes. We plan to build them in face diameters ranging from 14 inches to 60 inches, which is large enough to handle a log up to about 96 inches in diameter. Any reasonable carriage feed rate can be met with proper combinations of motor speed and number of knives to give a desired average chip length.

Motor sizes range from 75 up to possibly 1,000 horsepower for the large headrig chippers. Power requirements are related directly to the quantity of wood removed per unit of time. On some occasions, the power requirement may approach an astronomical figure. This can be compromised by slabbing in one, two or three stages. In other words, whittle the swell or the flare down to a normal slabbing position by short strokes.

A chipping edger is a worthwhile adjunct to the headrig chipper. It has many similar advantages in getting rid of the edging at its source. Addition of a top head can convert a chipping edger into a horizontal resaw with feed rates up to 600 feet per minute under some conditions. We believe that the utilization of new chipping equipment of this kind is a means of taking advantage of the greatest equipment development in sawmilling in many years. It is refreshing to realize that many mill operators are beginning to take advantage of this new equipment. I think it is becoming almost a necessity to save labor and utilize wood to the greatest advantage.

Finding horsepower

The chart in Figure 7.1 gives horsepower requirements, chipper-head revolutions per minute and number of knives. It assumes that average carriage feed rate, sectional area of slab and desired chip length are known. The carriage feed rate is determined by measurement when sawing logs of all sizes. It is modified to conform with the size from which the greatest chip production will come.

The sectional area is the cross section of slab to be removed; this section is taken, preferably, near the center of the log to represent the average. If large flares are to be chipped, this area should also be taken at the large end of the log. This figure should be used to determine the maximum horsepower requirement. If the maximum requirement does not exceed the average by 50 to

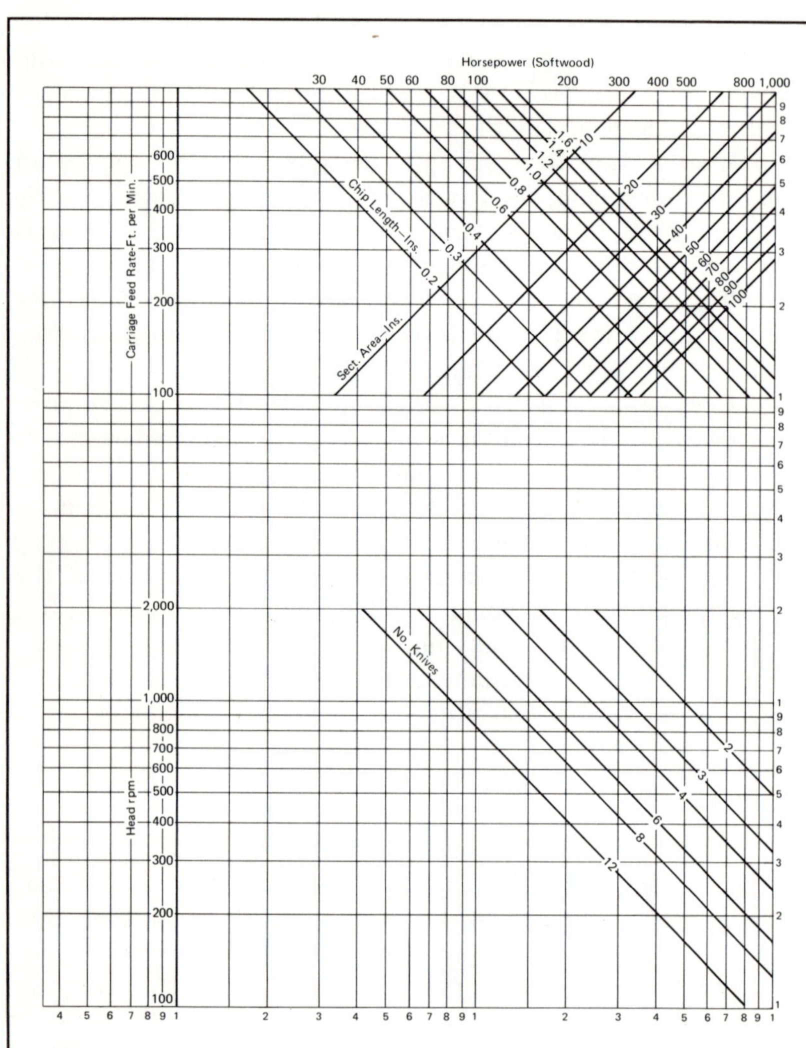

Figure 7.1. Chart for finding horsepower, suitable number of knives and chipping head revolutions per minute.

100 percent, the motor may tolerate this overload for short periods. Depth of cut can also be adjusted to limit overload on the motor.

The number of knives and head revolutions per minute are selected to give the best results for each particular application.

The following example will illustrate use of the chart (Figure 7.1). Feed rate, chip length and sectional area of the slab have been determined.

Feed rate	375.0 feet per minute
Chip length	0.8 inch
Sectional area of slab	20.0 square inches

Using these figures, the following information can be found from the chart (Figure 7.1).

Horsepower	250
Number of knives in the chipping head	8
Rpm of chipping head	700

Part 2 by Frank Roppel

Here are some experiences we at Ketchikan Spruce Mills, Inc. have had with our CM & E Model 64 headrig slabber, our employees' reactions and a few points other mill operators may wish to consider before installation of the equipment.

Production up 25 percent

First of all, we are quite pleased with the slabber and satisfied that it is doing the job we hoped it would do. At our Annette Island operation we cut primarily 8-inch and 4-inch cants for the Japan export market. Our average log is close to 22 inches. Accordingly, most logs would require five saw lines. By eliminating the slab cut, we eliminated two saw lines and increased our production at the headrig by 25 percent. Because of the relatively low number of lines per log, our production is chiefly controlled by the ability of the sawyer to load, turn and position logs on the carriage rather than by the carriage track speed.

We made some other changes at the same time the slabber was installed that also contributed to the production increase, but the slabber was the major item.

The slabber materially reduced the work of the offbearer. We still have the offbearer but at the end of the day he is not nearly as tired and there are fewer delays due to problems with slabs from flair butt logs. The edgerman's job is made easier by not having the slabs to work up.

Slabber cuts smooth

The smooth cut from the slabber is entirely acceptable to our customers. It is comparable to our bandsaw cut surface. Most of our production goes to remanufacturing plants; our customers have expressed satisfaction with the cut texture.

The quality of the slabber cut is not a problem to maintain and requires no special skills to set the eight knives in the head to the proper setting to maintain a smooth cut. It takes about 15 minutes to change the eight knives in our Model 64 slabber. The filer's helper changes the knives once each 16 hours of operation, usually between day and swing shift. We use a socket wrench but could do the job faster with an angle drive air or electric impact wrench, which we have not been able to find. (Perhaps one of you will design an air impact wrench that will do the job.)

Sawdust reduced—chips increased

The majority of our sawdust is, of course, produced on the headrig and edger. By eliminating sidecuts through use of the slabber, there was a noticeable reduction in the amount of sawdust and likewise an increase in chip recovery. This helped our burner situation and our chip income. Other chip increases were realized from slabs formerly sent to the burner because they were too wide for the chipper throat.

The filer maintains the slabber. The electrician maintains the setworks. Maintenance has been practically nothing on the slabber and very low on the setworks. Knives are ground in the filing room. It takes about five minutes per knife on our single-holder knife grinder.

We started with seven sets of knives in July 1972. Six of those seven sets are still in service. From our experience so far, it looks like we will go a year on those six sets. One of the initial sets was damaged by improper tightening without cleaning the holder. The knives will also break if not properly ground and released in the radius of the curve. It has been about four months since the last broken knife, so we think this problem is solved.

We have pulled logs off the carriage with the slabber and wrecked saws. The dogs were changed to a tong dog style and seemed to hold the logs better. Sawyer training and experience are critical to proper operation of the slabber, as one might expect, but the experience period is not as long as one might expect. The sawyers were all quite familiar with the slabber by the end of two days, fairly proficient at the end of two weeks, with proficiency increasing steadily for about two months at a noticeable rate.

Sawyer adjustment

When we installed the slabber, we had a serious problem with an older sawyer refusing to accept the concept of the equipment. We did not lay the proper groundwork to convince the man the equipment would help our operation. I think we finally convinced him it was a good improvement when we told him we were buying an air-conditioned cab with a slabber beneath. Once he became used to the slabber, he was our most enthusiastic supporter. However, to eliminate the problem of employee acceptance, I recommend the operators be familiarized with the advantages and utilization. Our young sawyer who was still breaking in was no problem. He adapted faster than the older, more experienced sawyer. A great deal depends on the individual. But if your sawyer is not sold on the equipment 100 percent, the equipment performance will not be 100 percent.

When you get ready to start using your slabber, its shadow lines are quite helpful. We started at the Annette operation without shadow lights or light beam markers. After installation of the lights, we found they were a real aid to the sawyer in full-speed utilization of the slabber. Good shadow lines aid sawyers in seeing where the slabber is going to cut so you do not overfeed the slabber.

And there is a limit to the depth of cut! Our Model 64 will take off a slab 6 inches thick.

The headrig slabbers are useful machines. One limitation is in frozen logs. They chip the log away, but the quality of the chip is not as high as in unfrozen logs. Normally a good quality chip is produced.

DISCUSSION PERIOD

MILTON MATER, Mater Engineering, Corvallis, Oregon: Does this slabber necessitate increased maintenance on the carriage and track? I am thinking of the pressures that the knives exert against the log, and the vibration. I wonder, perhaps, whether Frank Roppel has some information on that?

ROPPEL: We have not seen an increase in maintenance. You asked about vibration. Perhaps that is of more concern than maintenance. This machine cuts very smoothly. We were concerned at one time about the length of our chips being too short. One of the first things we did was remove four knives from our eight-knife machine. I do not recommend doing that because of vibration. We solved the length problem by replacing dogs and feeding faster. But we have not seen any increase in the maintenance of the track or in the carriage itself—no increase whatsoever.

WALTER COTTEN, Packwood Lumber Co., Packwood, Washington: What about twist? Do you have trouble holding the log and keeping a square cant as you are cutting?

MILLER: We have gone out of our way on some occasions to try to twist the log. If there is any twisting to be done, it will generally pull the dogs out of the log and the log will obviously be out of position. As long as the dogs remain fixed in the logs, I think there is very little danger of a twisted cant.

8

Increased productivity with headrig slabbers

Jim Bigelow, Bigelow Machinery Inc., Portland, Oregon

Cliff Jackson, Plant Engineer, American Forest Products Corp., Martell, California

Part 1 by Jim Bigelow

The Beloit Passavant headrig slabber is a multipurpose machine that will cure many ills because it will do many things. It will increase production up to three free lines per log. With little logs, two lines are saved and with big logs, four lines are saved. So, depending on lines saved, the percentage of production rises.

The headrig slabber eliminates swelled butts and in many mills does away with the offbearer. It will make possible reduction in kerf at the edger, as the edger will no longer have to break down swelled butts with heavy kerf saws. When the butts are too big for the chipper, they go to the edger if the chips are to be saved. Otherwise, they go to the hog and waste the chips.

Since the Beloit slabber makes whole log chips—cut out of the solid log—not wastewood chips, it will greatly increase the quality of the chips.

Upgrade chip mix

Whole log chips, added to ordinary chips, will upgrade quality. You could put the chips through unscreened. I made the mistake of telling one man that, however, and he took all the fans out of his blower. The chips must be screened because chunks will break off a rotten log at the headrig and fall down with the chips; so it is not the chips that need screening, but the other debris that also goes into the chute.

Occupational Safety & Health Act (OSHA) officials will be pleased. Another fellow we are going to help out is the sawyer. He will now be working in a soundproofed and air-conditioned cab with better visibility. Sound decibels will be reduced to 80, which is below the legal limit.

We have learned to let sawyers practice on other headrig slabbers before installation time. This acquaints them with the problems they will encounter when they get their own machines. You have to start with the man who is going to use the machine if you are going to make it run. These, then, are some of the multipurpose cures that installing a headrig slabber will provide.

There is no mill man, whether he is cutting 50,000 or 500,000 board feet, who should not take a hard look at a headrig slabber. One can be installed in almost any mill, over weekends, without downtime.

Beloit characteristics

When you decide on this big money-making cure, we think you should purchase the Beloit Passavant headrig slabber for the following reasons:

1. The Beloit is a one-piece, straight-knife slabber.
2. Knives are less costly to buy and easier to install than other types. Knives do not need special grinders; you may grind them on your own equipment.
3. You may buy knives for our machine from whoever sells knives. You are not wedded to us for knives.
4. The Beloit Passavant slabber has less carriage suck than other slabbers. I told this to a customer who bought one, but he forgot that fact. The day he started up the slabber, he said, "Everybody stand back. We're going to undock this log and let it go." But it just lay there at the slabber and ate up. There was no suck. So you can get by with less air on your carriage dogs. You do not need any more air; the regular

Beloit Passavant headrig slabber at American Forest Products Corp.'s Martell, California, plant. Note that the sawyer has the advantage of a wide field of vision. (Photo courtesy BELOIT CORPORATION.)

dogs will handle it. There is less pull and distortion at that point.

5. The Beloit is narrower in design and will usually fit into most saw boxes without costly moving of deck machinery, niggers and so forth. The slabber is only four feet wide. The normal saw box is four feet wide and we can get in there; we are designed for that.

6. The Beloit is more rugged. It has a larger drive shaft, bigger bearings and is built by the largest manufacturer of pulp and paper machinery in the world. Beloit knows how to make machinery to produce chips. Their reason for building this machine was to make more chips for their pulp and paper machinery. It is a total concept.

Setworks assets

All these advantages can be combined with the most accurate setworks. The C & D setworks, which sets metal to metal by rack and pawl, makes this the ideal combination for a real money-making machine. The C & D setworks is the only metal-to-metal setworks built that will set everything in your mill—carriage, edger, twin-band quad and slabber—accurately time after time, just as the old Trout did for many years. The full package costs about $30,000.

Hook this package into any computer. Wire it into the Western Union money counter at Mazatlan and take off. It will pump money down that line just as fast as you can put the logs by it.

With a double-cut headrig with fast-retract C & D setworks, every time the carriage moves there is a board. You are almost installing two headrigs with this slabber. On a double cut, there are different problems than with a single cut. The setworks is one of them. It must be very fast, very positive, like the C & D setworks. One mill operator reports a 17.2 percent production increase since his Beloit slabber was installed.

Sawyer preparation

In one mill that has a handyman and two regular sawyers, the men took to the slabber pretty well, but the company had sent

Beloit Passavant headrig slabber, Pickering Lumber B., Standard, California. Swelled butt is chipped ahead of the saw, saving one saw line.

Photo courtesy BELOIT CORPORATION

Right-hand, double-cut band with slabber combined with C & D rack and pawl setworks at Martell, California.   Photo courtesy BELOIT CORPORATION

160 MODERN SAWMILL TECHNIQUES

them to other mills for a day of indoctrination before they started to use the machine. Seeing the slabber work in other mills gave the men more confidence. There is no change in sawing speed. The sawyer is sawing to the saw, not the slabber. He is not concerned with the slabber, he is concerned with the saw. But there is money in every line—every line is a board, it is not a waste of time with a slab. It is a total management tool.

It is very easy to manage. Walk down the green chain. If you find a 10-inch chip face on a board, grab the nearest 2x4 and head for the saw box because there should not be anything down that chain over 4 inches wide or it is into chips.

Part 2 by Cliff Jackson

Our first concern at American Forest Products Corp. was to select a slabber that would fit into our existing space and, of course, our budget.

Setting plans and quotations were obtained from each manufacturer. Visits were made to various installations. The only slabber that would fit into the space allocated was the Beloit. The other manufactured slabbers, because of their rectangular shape, would have encroached upon our deck space at least an additional foot unless we had removed the supporting column to the filing room.

Mill criteria

After visiting various installations, the following criteria were established:

1. The sawyer's cab should be suspended from the floor of the filing room rather than supported from the floor up. This provides more room around the chipper and cuts down possible cab vibration from the saw.
2. Maximim visibility should be provided for the sawyer.
3. Nigger cylinders should be located farther away from the saw line, if possible, to prevent logs turning in front of the slabber head.
4. Control equipment should be mounted where it will receive minimum vibration and be accessible for maintenance.

5. A chain conveyor rather than a belt should be installed to insure positive chip removal.
6. A flop gate should be provided to divert chips to the hog conveyor in the event of chip-handling malfunctions either in or beyond the new conveyor we installed.
7. Simple, rapid means of changing, sharpening and locating chipper knife positions should be established.

By necessity, the entire installation at American Forest Products was accomplished on weekends, between other jobs. The net result was no lost time other than a final nigger adjustment to satisfy our sawyer's feel. This installation approach costs more, but our management could not justify a shutdown to install an auxiliary piece of equipment.

The sawyer's cab was suspended from the floor joists of the filing room, distributing the load over several joists. The floor of the cab was located 6 inches above the chipperhead, with the front of the cab 20 inches from the saw line.

It was a double-cut headrig installation. To provide maximum visibility for the sawyer, a section of the top bandmill wheel housing was removed, the electric feed pot was reversed to place the pot away from the saw line and a new sawyer's control console was suspended overhead in front of the sawyer.

Control locations

The new control console contained all functions previously located in several consoles, but in their relative positions. This eliminates the need for the sawyer to learn new switch positions. A television camera was installed 30 feet down the rollcase behind the bandmill to give the sawyer a view of the bumpers to the resaw, edger, floor chains and the camelback at the end of the rollcase. That left one blind spot directly behind the bandmill, which was remedied by installing a mirror directly across the tracks.

An unforeseen situation developed immediately after installing the slabber. When the sawyer finished his cut on the cant with a forward motion, he could actuate the slabber setworks as he dropped off the cant. But if he was cutting toward the deck, he had both hands full and had to wait for the cant to clear the chipper and have the next log completely loaded and dogged. This resulted in a partially missed line while waiting for the chipper to come into position. To overcome this, we installed a single button in the nigger handle—it could have been a foot switch—that, when actuated, either moves the slabber to a preset cutting position or retracts the slabber out of the way. This function can be performed automatically, but we do not feel it would enhance our operation at all.

Chip conveyor

A single H110 chip conveyor 13 inches wide by 12 inches deep was installed. It travels at 60 feet per minute to convey chips from the slabber. This conveyor was purposely extended to the single-cut (headrig) side to provide future chip removal when the second slabber is installed. (I am optimistic, hoping that Bendix will buy another one.) A simple flop gate was installed to divert chips to the hog conveyor to save possible future downtime.

Straight knives on this slabber can be changed, with knives pre-set to a gauge, in 15 minutes or less by a person who has never done it. The only tools used to change knives are an Allen wrench and a magnetic gauge. With these two simple tools, knives can be changed in about 10 minutes. An impact wrench would be a handy tool for changing knives, but it is not absolutely necessary.

At first, the knives were ground on a standard grinder by an operator who was not properly instructed. As a result, the face portion of the knife came out with varying angles and produced a surface that did not have the best appearance. A simple gauge was made for the grinder operator and subsequent grinding produced proper knife angles. Slab surface appearance improved. A finer finish can be easily and quickly provided by jointing these straight knife surfaces. Our men modified a jointer that had been discarded from the planing mill to fit the chipper.

Slabber results

The slabber has increased production in our operation by direct-ly increasing the number of usable lines sawn and eliminating time formerly lost developing slabs. It has put the slabs where they belong—into usable chips.

American Forest Products Corp., Amador-Calaveras division, a Bendix enterprise, would welcome industry visitors to our facility. We feel certain that the visitor would learn from observation. At the same time, we could acquire information from him to enhance our operation.

DISCUSSION PERIOD

VERNON S. WHITE, editor, Western Timber Industry: After you put that video operation into your sawyer's cab, how did you stop him from watching soap operas all morning? This is a half-way serious question, Cliff.

JACKSON: Another recipient of a Beloit Passavant chipper had a video section television receiver installed in his cab and the electrician hooked it up for the local ball games. You may think this is a joke, but that is what he did! The manager put a stop to that! We have, instead of the ball games, what we call the white fir parade. These are logs that are the rejects from the plywood mill; they are the crummiest things you can possibly get. We would like to have cut all the pine if we could; we do cut quite a bit.

MILTON MATER, Mater Engineering, Corvallis, Oregon: When tying the slabber into your carriage track and frame, what did you run into? Did you go right down to the ground or did you tie into your columns? And did you tie it across to your V-rail?

JACKSON: Actually, the slabber is tied into the substructure that is also the substructure for the V and flat rail. All we did was extend the foundation, or I should say substructure, approximately 6 feet. If you have enough lead in your head so that the back portion of your knives is not dragging, you have no pull or push on the carriage. We did, however, experience this. We had a negative lead in that head to start. Believe me, it will push the knees. But actually your knife is only sticking out about 1/32 inch; if you do not have the lead in the head, you are going to have to push because the cant is going to hit the face. We corrected that quickly—we twisted the chipperhead.

MATER: Did you tie it down to the ground at all?

JACKSON: No, it is already tied with your columnar structure in the mill, which is tied down to the ground, so there is not that much inertia. There is no tendency to tear the rails out of place. You can put it on top of your present structure.

RAY SWANSON, Swanson Bros. Lumber Co., Inc., Noti, Oregon: I have two questions: (1) What is the maximum width of the face that you can get with a Beloit Passavant chipper? and (2) How far can you go in making long chips in the setting of your knives and gauging the speed of your machine?

JACKSON: Well, I do not know. Beloit has made, or has in production, heads of almost any size. At American Forest Products we have a six-knife head, 24 inches from point to point, and we have dropped the bottom of the head down one inch below the carriage ways. This gives us 23 inches above the knees. Changing speeds is easy by changing sheaves. This is the only way you are going to do it because it is a function of the number of knives that are hitting and the rate of speed. As to rate of speed, this machine has ample horsepower. We have a 200-horsepower motor on it. At present, with sharp knives, we draw as little as 70 amperes, which, as you know, is ridiculous with a 200-horse motor. However, we have run our knives for as much as three nine-hour shifts. As they get duller and duller, the surface deteriorates and horsepower goes up.

SWANSON: My question was how far can you go in getting a long chip?

JACKSON: This depends on your sawyer. It is a function of feed speed. We are producing a chip in excess of 5/8 inch long. That is what we are striving for, but at times our sawyer produces chips more than an inch long. It depends on how hard he chops that gun back and goes into the cut.

SWANSON: The reason for my question is that our customers want chips 7/8 to 1 1/8 inches long.

JACKSON: You can get them. Believe me, you can get them. Our sawyer is producing them and we do not want them.

CHUCK MURRAY, ITT-Rayonier, Shelton, Washington: This is not so much a question as a comment. This is the last paper on chipper slabbers, or chip-producing machines. Much has been said about using these machines to improve lumber production and get more chips. But nothing really has been said about the quality of the chips. I think we ought to be aware that there are going to be some changes from the pulp mills as to what they want from the sawmills in chips. At a TAPPI (Technical Association of the Pulp & Paper Industry) meeting in Tacoma, an Oregon kraft mill announced that it had applied certain changes in chip requirements and managed to save $126,000 in less than a year by tightening up chip specifications. Are there comments from other sawmill operators as to what is going on between them and their customers?

JACKSON: Well, sir, we had the representatives of Fibreboard Corp., one of our customers, take a look at our chips. They were perfectly happy with them. It may be that they are going to tighten restrictions, but I do not know for sure what you are talking about. I imagine you are talking about screening, and this can be accomplished. We can screen chips differently if they desire it. But our chip meets their specifications because it is a whole log chip, rather than a bobbling end hitting the chipper and producing what I call a bone or a long piece. These chips are of relatively uniform length because it is a function of the carriage speed.

9

High-strain/thin kerf

F. E. "Ed" Allen, Chief Engineer, Letson & Burpee, Ltd. Vancouver, British Columbia

I think it is necessary to give a brief history of how high-strain/ thin kerf started. Telling you about the theory will show that it is not black magic, although the birth of the theory at Letson & Burpee, Ltd. was the result of an accident.

In 1965 we received an order for a 7-foot bandmill from a company in Quebec. Although we had manufactured Monarch bandmills under license for many years, we had not been overly successful financially. With the new order, we decided to do a complete redesign and introduce a new bandmill model. We had two things in mind: (1) to improve the model and (2) to manufacture it more economically. We completed the manufacture and prepared the mill for shipment. For some reason, which I do not remember now, we were required to ship with the upper wheel in position. To protect the delicate knife-edge strain system on a 3,500-mile rail journey, we installed special blocks to absorb vibration during shipping. The blocks were painted bright yellow to contrast with our normal machinery colors. In our instruction manual, the reason for the blocks was explained and instructions were given for their removal before startup. But in the confusion that frequently accompanies such operations, no one bothered to read the directions. The machine was placed in operation and, from telephone conversations, seemed to be performing exceptionally well. Within a few months, however, an arbor failure was reported. I did some fatigue calculations that showed there was no way the arbor should fail, so I decided it must have been machined from a faulty forging. We sent a new arbor and paid to have it installed as a condition of warranty.

After about the same period of time, there was a second arbor failure. So when we shipped another new arbor, we sent a man along to see what was happening. When he arrived, the first thing he saw was the bright yellow blocks still in position on the machine. On questioning procedures, he found that the strain was applied by pressing the raise button for the wheel lift motor. When the motor stopped, the man took his finger off the button and that was the operating strain.

This time, with the third arbor in position and our bright yellow blocks removed, production started, but not very successfully. It was midwinter. Logs were frozen and the machine was working under conventional strain. No amount of care would stop the saw from snaking all over the place.

It seems that whatever strain we actually had on that mill somehow gave the operator the capability of sawing accurately and at acceptable feed speeds.

Faux pas initiated study

The faux pas at installation was our first involvement in a concept we now call high-strain. It initiated a long and continuing study on bandmills and bandsaws. This study provided a basis for design. Because our first problems with high-strain were fatigue failures, it was reasonable that we borrow knowledge developed by aircraft and automotive engineers in fatigue design; with this knowledge we could redesign our bandmills and determine safe working stress for bandsaws with strain applied.

We first had to determine the fatigue strength of bandsaw steel and balance it against all stresses involved. When we had done this, we could establish the design strain for each size bandmill.

Our first attempts at a modified Goodman diagram were very theoretical, but we were fortunate in prevailing on Sandvik of Sweden to do the necessary tests for us so that a correct Goodman diagram could be established (Figure 9.1). These tests gave the released loading fatigue strength of plain and of notched specimens. The stress concentration factor was almost identical with that indicated from photoelastic studies.* It is important to appreciate the effect of bending on a bandsaw blade. By taking the heavy line starting at 30,000 pounds per square inch (Figure 9.1) and going upward, any place in between those heavy lines is a safe area to work in for fluctuating stress. I think nearly everyone has taken a piece of wire and bent it to break it. The same thing happens with a bandsaw—the fluctuating stress going around the

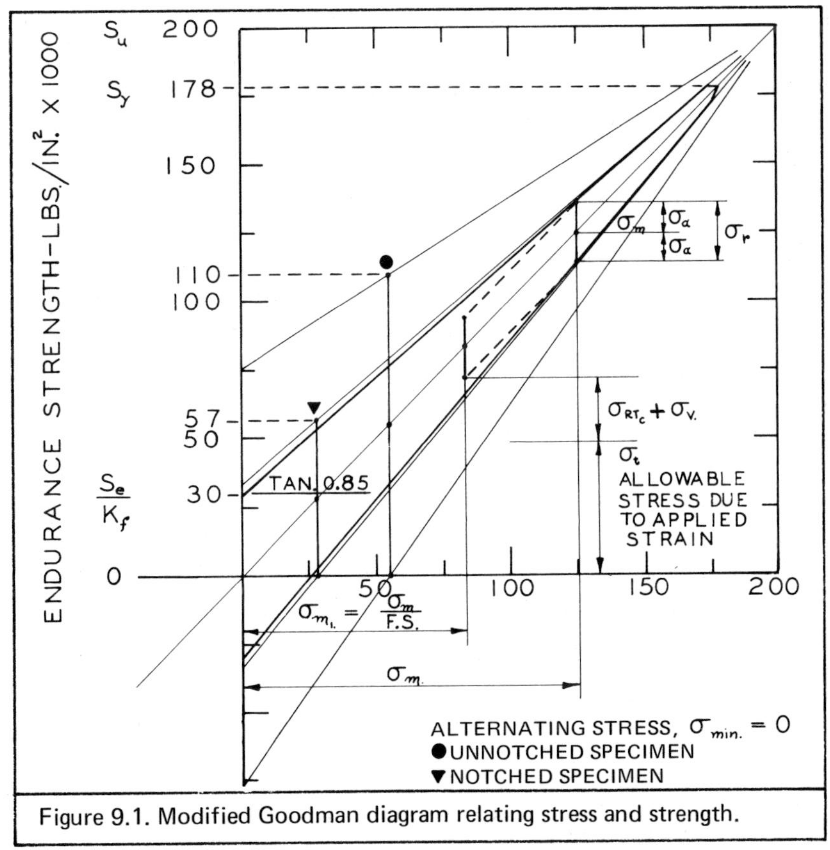

Figure 9.1. Modified Goodman diagram relating stress and strength.

* Allen, F. E. "The Merits of the High-Strain Bandmill." *Forest Industries Review* (New Zealand), December 1972.

wheel and then straightening between wheels is the primary cause of fatigue cracks.

This was the beginning. The obvious soon became apparent—the fluctuating stress is the bending stress. To work safely to a higher stress and to a higher strain we must reduce the fluctuating stress. That really means using a thinner saw. The practical limitation of this statement is a saw thickness of approximately 1/1500 of the wheel diameter. Beyond this ratio, the area reduces more rapidly than the bending stress and the allowable strain starts to decrease. The range stress, shown in Figure 9.1, is simply the bending stress corrected to include effects of anticlastic curvature and chord height. The bending stress may be calculated by multiplying the modulus of elasticity by the ratio of the plate thickness, divided by the circle's diameter in which the saw is bent.

Anticlastic curvature is undoubtedly the sawfiler's worst enemy in tensioning thin saws for use with high-strain. The term *anticlastic curvature* is not really confusing. If we take a rubber eraser and bend it, the outside edge, because it is in tension, decreases in width. The inside, because it is in compression, increases in width. This causes a curve in the transverse plane. This also happens to a bandsaw if it is incorrectly roll-tensioned, or if too wide a tire line is left. In checking this at the diameter of the wheel on which it is to be run, you will frequently find that the edges tend to curl.

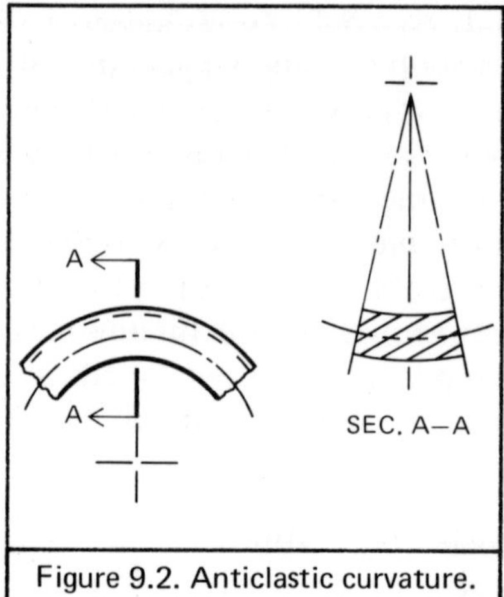

Figure 9.2. Anticlastic curvature.

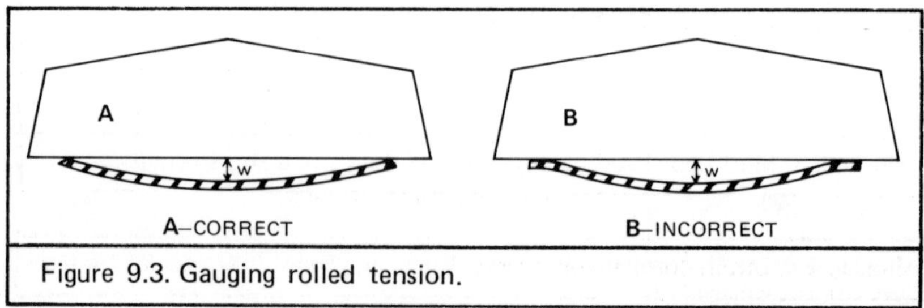

Figure 9.3. Gauging rolled tension.

Figure 9.2 shows a cross section and the effect of anticlastic curvature on any piece of metal, or any substance that is bent.

In Figure 9.3, the top drawing (A) shows the correct cross-sectional appearance of a bandsaw when it is tensioned and gauged at the wheel diameter on which it is going to be run. The lower portion (B) shows a saw that is roll-tensioned and gauged in the conventional manner; if it is rechecked at the wheel diameter, it frequently appears as shown. (Both figures have been formed into circles equal to the wheel diameter.) To operate satisfactorily, the bandsaw must be roll-tensioned to overcome anticlastic curvature.

High-strain defined

What is high-strain? Is it twice conventional strain, three times, or how much? The answer is, of course, that there is no relationship between conventional strain and high-strain. Conventional strain is based on a constant stress of 5,000 pounds per square

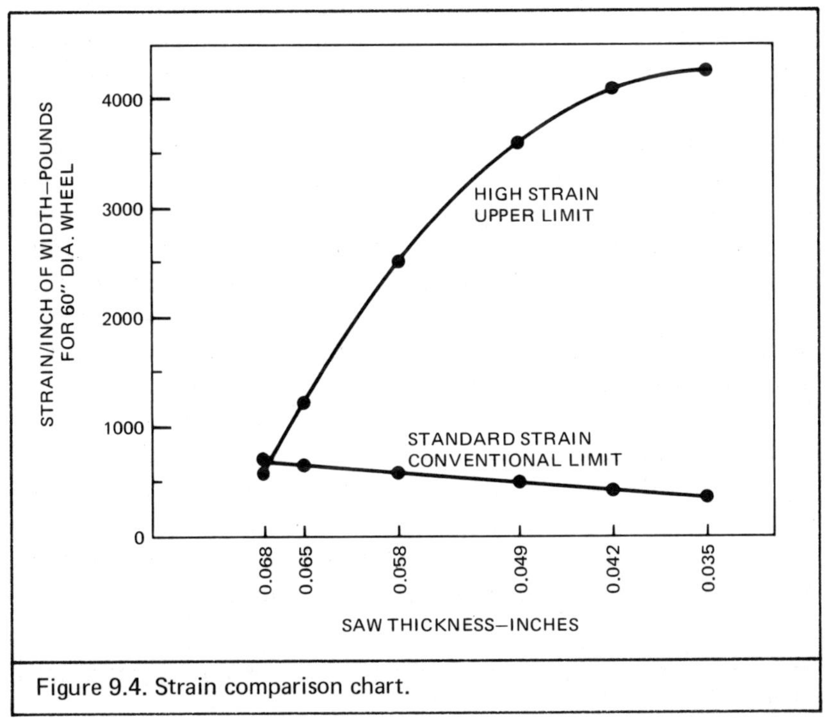

Figure 9.4. Strain comparison chart.

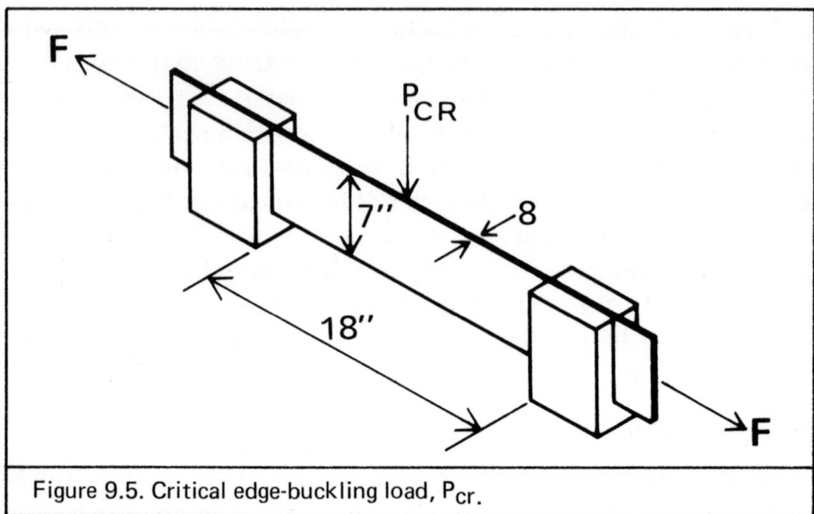

Figure 9.5. Critical edge-buckling load, P_{cr}.

inch. High-strain is based on the principle of balancing stress and strength, as shown in the Goodman diagram, Figure 9.1.

Perhaps the best comparison is to show the working strain—as determined by high-strain equations and by conventional strain equations—for several saw gauges that might be used on a 60-inch bandmill.

In Figure 9.4, the bottom line on the graph shows conventional strain limits for saws of different gauges. The upper curve shows strain limits for the same saw gauges when high-strain equations are applied. The diagram shows that there is obviously no relationship between the two. In a fairly heavy saw, high-strain equations would actually give a lower value than that given by conventional strain equations. Thin saws and high-strain run hand in hand. One will not work without the other.* As the saw plate is reduced, the allowable stress increases and thus allowable strain increases.

Between 1968 and 1970, two groups independently attacked the problem of quantifying the theoretical and practical improvements that could be obtained through high-strain.

Research scientists at the Vancouver Forest Products Laboratory published an information report on "Lateral & Edge Stability

* Jones, D. S. "Gullet Cracking in Saws." *The Australian Timber Journal* no. 2 (August 1965).

of High-Strain Bandsaws."* Letson & Burpee, Ltd. started a series of practical tests, which are still in progress, to determine what high-strain can do for the end user. The critical edge-buckling load equations, published in the Forest Products Laboratory report, indicate that a saw of given thickness, width and distance between guides has a critical buckling load (Figure 9.5).

If we consider the saw simply as a beam that is supported and held from falling over (without any additional force applied) it has some critical buckling load due to the moment of inertia of the section. A tension force increases this critical edge-buckling load proportionately.

Figure 9.6 shows several gauges of saws that might be used with

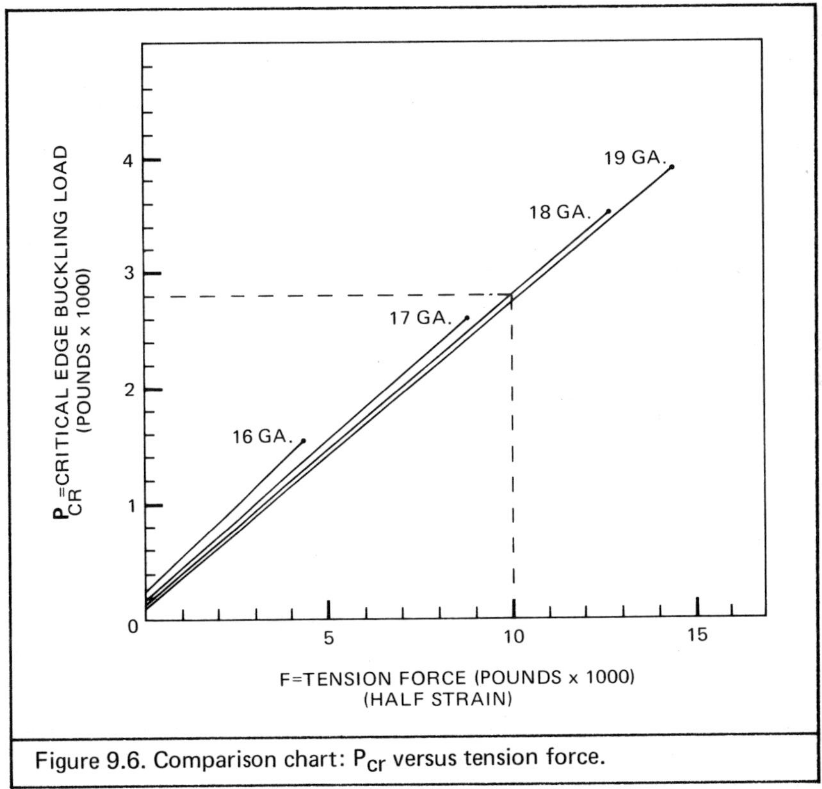

Figure 9.6. Comparison chart: P_{cr} versus tension force.

* Foschi, R. O., and Porter, A. W. "Lateral and Edge Stability of High-Strain Bandsaws." *Vancouver Forest Products Laboratory Information Report*, 1970.

a 60-inch-diameter wheel. The horizontal axis shows the tension that might be applied (this is half strain) and the vertical axis, the critical edge-buckling load that might be obtained from each saw with different applied tension forces. We see that because of the limited strain a 16-gauge saw can safely carry it has a certain critical buckling load. We also see that we are able to carry a much higher tension force by going to a thinner saw, thus producing a considerably higher critical edge-buckling load. A 16-gauge saw 7 inches wide can carry a tension force to 4,350 pounds, which produces a critical edge-buckling load of 1,550 pounds. By reducing thickness, we can increase the tension force and thus increase the critical edge-buckling load. For example, if F = 10,000 pounds, the critical edge-buckling load on an 18-gauge saw becomes 2,820 pounds. Theoretically, at least, this should improve our bandsawing capabilities tremendously.

But there is much more to making a bandsaw function properly than improving its critical edge-buckling load. There are three other areas of equal importance. Without all four, there will be a problem that will be evident in the output of bandmills. More clearly, there is one area over which we have absolutely no control. That is the characteristic of the material that we are machining. The bandsaw is generally considered to be quite a simple and straightforward piece of equipment. Believe me, there is nothing further from the truth.

The bandsaw is as sensitive and as easily hurt as any fair maiden. It is the responsibility of machinery manufacturers to make sure their equipment is capable of giving maximum protection to the bandsaw.

The prime function of a bandmill strain system is to protect the bandsaw from being hurt. If something happens in the cut, the strain system must respond instantly and efficiently and the corrective action must dampen out as quickly as possible.

We have developed tests for measuring response time, by recording on a dual trace-memory oscilloscope the time lag in milliseconds between a severe disturbance of the blade and the corresponding reaction by the strain system. It is important that this reaction time be no greater than 10 milliseconds, which is one-hundredth of a second. It is more important that this happen without significant change in stress in the bandsaw blade.

Hysteresis energy loss

To analyze the efficiency of a strain system, we mount strain gauges on the saw blade at the back of the bandmill so we can read directly in microstrain. One microstrain represents a stress of 30 pounds per square inch.

Whenever a certain amount of additional strain energy is added to a stressed bandsaw blade and then removed, only a fraction of the input energy is recovered. The remainder is lost in friction and is known as hysteresis energy.

To measure this energy loss, a reading is taken on the strain indicator with the blade in its neutral position. Then the blade is displaced accurately in 1/10-inch increments up to 0.8 inch and a reading recorded at each position. Displacement of the blade from zero position represents additional energy being put into the system. The reverse procedure indicates strain energy leaving the system. If these points are plotted on a graph, with displacement on the X axis and the Y axis indicating the microstrain, then the area within the curve thus formed results is a measure of the overall

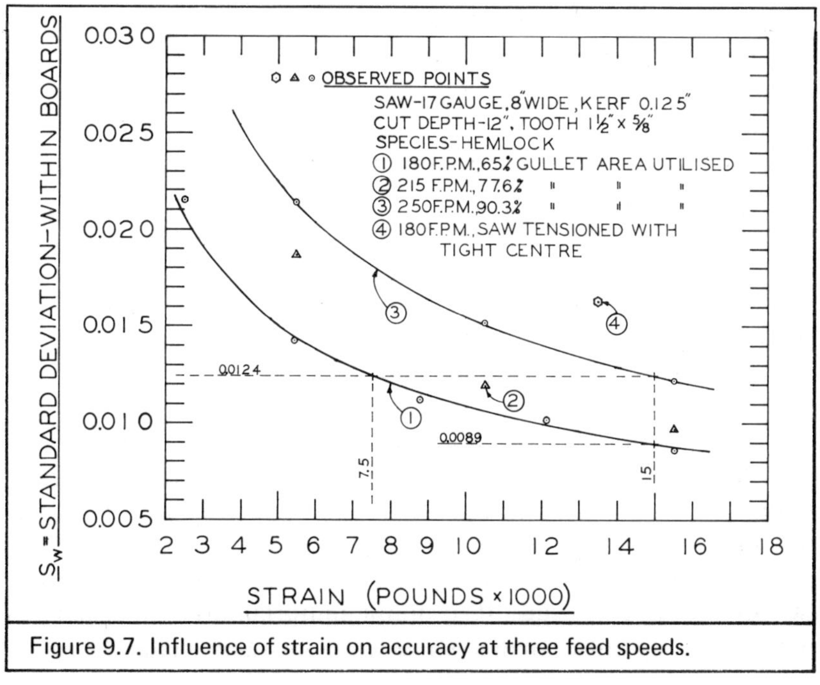

Figure 9.7. Influence of strain on accuracy at three feed speeds.

efficiency of a bandmill strain system.

The ability to measure what our equipment is doing and put numbers to it has allowed us to improve the speed of response to 35 percent of that found in our tests of 1971. The hysteresis energy loss has been reduced to 27 percent of our 1971 values and the sawing accuracy in our recent saw trials has been improved by 25.2 percent.

Strain influences accuracy

Strain influences accuracy. To the sawmill operator results, not theory, are a basis for judgment. To practicing engineers, results can only be improved by refinement in theory and improvement in equipment. During December 1972 and January and February 1973, we conducted sawing tests at our Surrey, British Columbia plant. These tests were performed on our prototype, full air strain bandmill. To date, more than 10,000 measurements have been made to gauge performance.

Figure 9.7 has been compiled to show the influence of strain on accuracy and the influence of increased utilization of available gullet area on accuracy. Figure 9.7 represents 12 sawing tests of 50 boards each, with a total of 7,200 measurements.

Opponents of the theory of high-strain and thin kerf have frequently expounded their belief that for this concept to work it is necessary to saw at substantially reduced speeds. To show the fallacy of this statement, I have chosen two strain levels—7,500 and 15,000 pounds. From the 7,500-pound strain level, a vertical line is projected to intersect curve number 1 at a standard deviation of 0.0124 inch (Figure 9.7). Curve number 1 used 65 percent of the gullet area for removal of solid wood.

When the strain is doubled (from 7,500 to 15,000 pounds), the operator appears to have two alternatives: (1) remain at 65 percent level and improve accuracy to 0.0089 inch, or (2) remain at the same accuracy level and increase feed speed to 90 percent utilization of the gullet. The first alternative represents an improvement of 28.2 percent in accuracy by doubling the strain. The second represents an increase in feed speed of 38.9 percent while maintaining accuracy—again by doubling the strain.

Tests represented on this graph were conducted with 17-gauge saws so that a wide range of strain levels—2,500 to 15,500 pounds—could be used to indicate more clearly the effects of strain on accuracy.

Figure 9.8, a computer printout, shows a 19-gauge saw run at 24,000 pounds strain and 0.095-inch kerf. The saw is 7½ inches wide, has 1½-inch pitch, 5/8-inch gullet depth, used at 180 feet per minute in a 12-inch cut. Standard deviation on a 50 board test is 0.0086 inch within boards and 0.0068-inch standard deviation between boards.

Accuracy of this test was equal to the best we were getting from the 17-gauge saw and a kerf saving of 0.030 inch was obtained. The left-hand column indicates board numbers; the next column is the number of measurements taken in each board; the third column is the range of sizes within each board; the fourth column shows the average size; the last column is the standard deviation.

When we planned these tests (using a 60-inch bandmill on slides with screw adjustment, no setworks, and used with a fixed line-bar), we arbitrarily set the board size so as to be able to get 10 pieces out of a 12-inch cant with the two outside boards being thrown away. This meant subtracting 11 times the kerf width from 12 inches and dividing the remainder by 12, which gave a size approximately 7/8-inch thick. By putting five cants through per test, we automatically obtained our 50 boards.

Same patterns with 19-gauge saw

With the 19-gauge saw, we followed the exact same patterns, took out the interchangeable saw guides and replaced them with a set that had been remachined. When we ran the cants through, to my astonishment, we ended up with 11 boards (one extra) from each cant. The outside board, instead of being a full 7/8-inch thick, ended up as a 3/8-inch shim. One of our young chaps thought this was a pretty sneaky idea so he took two more cants. He sawed out one with a 17-gauge saw and the other with a 19-gauge saw. Then, he put all the boards together and cut them so he could stack them together and show the difference in height between the two. There was a full 3/8-inch difference.

We did another test that will be of interest because it concerned the feed speed used. We used a 17-gauge saw, 7½ inches wide,

Calculation of Average Thickness, Range, and Standard Deviation Within and Between Boards

Performance Testing of New Letson & Burpee Air Strain Bandmill   Jan. 24 / 73

Actual Strain = 24,000#   Test #9   Feed Rate = 180 FPM

19 Gauge   1.5" Pitch,   5/8" Gullet Depth

Board No.	No. of Readings	Range (Ins.)	Mean (Ins.)	SD Within (Ins.)
2001	12	0.026	0.8814	0.0096
2002	12	0.016	0.9010	0.0057
2003	12	0.040	0.8947	0.0107
2004	12	0.014	0.9022	0.0047
2005	12	0.031	0.8972	0.0101
2006	12	0.038	0.8843	0.0137
2007	12	0.022	0.9049	0.0062
2008	12	0.040	0.8974	0.0103
2009	12	0.021	0.9041	0.0058
2010	12	0.024	0.9032	0.0074
2011	12	0.025	0.8958	0.0077
2012	12	0.050	0.9037	0.0138
2013	12	0.038	0.8952	0.0130
2014	12	0.014	0.9087	0.0052
2015	12	0.021	0.9017	0.0055
2016	12	0.015	0.9002	0.0041
2017	12	0.039	0.9037	0.0102
2018	12	0.039	0.8986	0.0107
2019	12	0.009	0.9059	0.0033
2020	12	0.012	0.9050	0.0046
2021	12	0.045	0.8890	0.0129
2022	12	0.040	0.9011	0.0105
2023	12	0.024	0.9011	0.0074
2024	12	0.017	0.9077	0.0052
2025	12	0.018	0.9060	0.0052
2026	12	0.048	0.8976	0.0140
2027	12	0.030	0.9034	0.0086
2028	12	0.022	0.9030	0.0071
2029	12	0.024	0.9109	0.0070
2030	12	0.023	0.9084	0.0063
2031	12	0.048	0.8956	0.0154
2032	12	0.027	0.9008	0.0092
2033	12	0.012	0.9042	0.0043
2034	12	0.037	0.9057	0.0114
2035	12	0.024	0.9107	0.0072
2036	12	0.008	0.9023	0.0029
2037	12	0.034	0.8947	0.0111
2038	12	0.024	0.9016	0.0065
2039	12	0.027	0.9012	0.0103
2040	12	0.022	0.9092	0.0067
2041	12	0.028	0.9026	0.0084
2042	12	0.016	0.9017	0.0059
2043	12	0.016	0.9088	0.0069
2044	12	0.028	0.8984	0.0083
2045	12	0.014	0.9071	0.0043
2046	12	0.018	0.8995	0.0051
2047	12	0.044	0.8941	0.0151
2048	12	0.011	0.9138	0.0041
2049	12	0.025	0.9002	0.0075
2050	12	0.017	0.9118	0.0060

Grand Mean Thickness (Ins.)   0.9016

Mean Range in Thickness (Ins.) 0.0261

Range of Mean Thickness (Ins.) 0.0324

Pooled Standard Deviation Within Boards (Ins.)   0.0086

Standard Deviation Between Boards (Ins.)   0.0068

Number of Boards   50

Figure 9.8. Performance test results of new Letson & Burpee air strain bandmill. Computer printout shows a 19-gauge saw run at 24,000 pounds strain and 0.095-inch kerf.

gullet to back, 1¾-inch tooth pitch, ¾-inch gullet depth. Gullet area was 0.803 square inch. Feed speed was 325 feet per minute. This represents a bite per tooth of

$$t = \frac{F \times P}{C} = \frac{325 \times 1¾}{8560} = 0.0664 \text{ inch}$$

The area of solid wood removed with each tooth equals

$$V = t \times (D\text{-}0.5P) = 0.0664 \times (12\text{-}7/8) = 0.7391 \text{ inch}^2$$

Taken as a percentage of the total gullet area, it equals

$$\frac{0.7391}{0.803} \times 100 = 92.05\%$$

Considering the degree that this skinny little saw was overfed, the decay in accuracy was less than would have been expected. Standard deviation within boards was 0.0186 inch. Standard deviation between boards was 0.0130 inch.

Consider these few points on bandsaws: Because the actual stresses involved in the operation of a bandsaw must be known very closely for high-strain equations, we have delved into this aspect very diligently. We can calculate the amount of prestress that is desirable in a saw, but we were never quite sure whether we actually had that amount or whether we had correct distribution. Saws used in our tests were formed into a 30-inch radius and the light slit was gauged electronically so we knew exactly what the distribution of rolled tension was like.

We had Fred Clark, our saw engineer, change the tension of shapes that we have frequently seen—tight quarters, tight centers—and gauged these so we had printout records of their shapes. Figure 9.7, point number 4, was a saw with a tight center. It was run at 65 percent gullet utilization. The standard deviation within boards was 0.0163 inch compared to 0.0086 inch with a correctly tensioned saw. Our tests have proved conclusively the benefits that can be obtained through the use of high-strain in conjunction with good saw filing techniques and the most responsive bandmill on the market today.

[Allen then narrated slides. A 19-gauge saw was shown at different feed speeds. The first was feeding at 100 feet per minute with sawdust visible still in the gullet. The second showed the same saw used at 180 feet per minute. The saw had 1½-inch pitch, 5/8-inch gullet, 0.095-inch kerf. The third showed the saw speed at 220 feet per minute. The fourth showed the saw cutting at 275 feet per minute in a 12-inch-deep cut. It was grossly overfed as evidenced by the explosion of sawdust in all directions.]

The Letson & Burpee bandmill is a full air strain mill. The throat of the mill has been completely cleared out. We have 7 inches more throat than on our previous model. There are no push rods, no column tubes or anything coming down in front of the mill. Strain is applied by a frictionless diaphragm-type cylinder back of the wheel. The strain cylinder has stops that limit the lower and upper travel of the strain rocker arm.

Guide blocks are removable. You loosen one bolt, take the blocks out to the filing room and put back a new set that has been remachined. The block can be set out slightly in its position in the jig and refaced. Once this is done precisely, it is not necessary for the filer to adjust the saw guides each time he changes the saw. He can change the guides every four hours or every six hours or every eight hours. Whenever he changes saws, he can take the two guides off. They are completely interchangeable. It does not matter whether it is a twin or a quad; the guides go on number 1 mill and on number 4, top or bottom.

DISCUSSION PERIOD

MILTON MATER, Mater Engineering, Corvallis, Oregon: Would you discuss your definition of standard deviation—whether it is plus or minus, how you arrived at it and what measuring devices were used?

ALLEN: Standard deviation is a means of taking measurements and using them so that they are statistically significant. We can take measurements with dial calipers and we get a range within boards; but if we just do this on 50 boards, it becomes a meaningless function. Basically, what we do is measure each board 12 times with dial calipers that measure in thousandths of an inch. We feed this information into the computer and get what we call standard deviation within boards so we can compare our sawing accuracy. We get standard deviation between boards so we can ascertain whether our feed system is in poor shape or is moving. We can combine this to find out what the standard deviation of our total production is. Now this standard deviation curve represents total production. The center line in the bar X represents the mean size. One standard deviation on either side of the mean size encompasses 68 percent of total production; two standard deviations include 95 percent of total production.

It is normal to take a lumber size. We have a minimum value we can live with that will clean up in the planer. This includes surface finish and everything else. On top of that, we must add two standard deviations to take care of sawing accuracy. This will provide for a 2.13 percent falldown, and that will give us a means of measuring and having quality control on our end product. This same system is used entirely in the automobile industry, and it is also used now by some of the leading forest industries, both in Canada and the United States. Weyerhaeuser, for one, is using it.

MATER: Would you discuss the units you used and whether we are talking about plus or minus?

ALLEN: We measure in thousandths of an inch. We are still rather primitive; we have not changed to the metric system yet. Any board we would measure would have a mean size. If we added up all the measurements and divided by the number of measurements we had taken, we would have a mean size or an average size. Of measurements in that board, then, 68 percent would be within plus or minus one standard deviation; 95 percent of the measurements within that board would be within plus or minus two standard deviations. In an individual board this is really not all that meaningful. But because of machines and the characteristics in the wood itself, if we take 50 boards we have 50 different mean sizes that are distributed over a slight range. We take standard deviation generally within each board, within boards for a group and between mean sizes. If the standard deviation *between* boards is more than about three-quarters of the standard deviation within boards, this would indicate very strongly that we had a problem in our feed system, carriages or whatever it might be—a setworks problem. This is one reason that we use the two alternatives, within and between, and we do measure in thousandths of an inch with dial calipers.

MAGNUS GUNDERSON, Simpson Timber Co., Hudson Bay, Saskatchewan: In one of our bandmills, one of our bearings was running very warm. They were concerned about it on the bottom. How warm can they run?

ALLEN: Well, if you are running fairly high-strain, bearings will normally run up to 120 to 140 degrees without your having to worry.

GUNDERSON: There was a change in your manual on greasing, I think, at one time. This was put in about two years ago.

ALLEN: We changed the type of grease because we found we were getting a breakdown of oil separating from the main body of the grease. We switched to another grade, not another make. You should have no trouble with the bearings if they are running hot. They do create a lot of energy because of deformation in the rollers and the races. This energy has no other form to take but that of heat.

10

Tradeoffs at the headrig: production vs. recovery

P.S. "Phil" Quelch, West Vancouver, British Columbia

Today there is a great deal of talk in the industry about thin kerf on headrigs. This talk is in the right direction; anytime the industry does something to save the forests, there certainly can be no complaint.

My question is this: Do we keep our priorities in the sawmills in proper sequence? Keeping in mind that we are trying to save a few thousandths of an inch on a headrig kerf, we go down to the edger and see the flagrant waste that takes place many times from over-edging to establish stock from random-width cants.

Let us take a popular favorite of today, the thin kerf bandsaw. Let me state at the beginning that there is no way to cut as much lumber with a thin saw as with a heavy saw, regardless of high-strain or other gimmicks. It is for management to decide if the operation will be more profitable with a thin kerf and less production than with a heavy saw and high production. But do not be led into believing you can have the best of two worlds.

Next I ask: If you have reduced the kerf, what are you going to do with it, unless you have the latest in setworks with the built-in saw kerf? Let us take a very simple example. The sawyer wants to finish up with 4 inches on the head blocks and has enough stock to make two pieces of 2-inch, so allowing two saw cuts of ¼ inch each, he will set the dial on 8½ inches to slab, 6¼ on the second cut and 4 inches on the third line leaving 4 inches on the blocks. Now by going to a 17-gauge saw from a 13-gauge we have saved approximately 0.058 inch. To take advantage of this saving the sawyer must now do some fast figuring. First of all, he must consider that the 13-gauge saw did not take a ¼-inch kerf which was allowed but only 0.200. A 17-gauge saw 12 inches wide would take 0.142, so we have saved 0.058. You tell me how a sawyer is to calculate in thousandths of an inch. Even if he could, most setworks are just not that accurate.

Trimmers wasteful

And if you think that the waste at the edger is bad, walk on down to the trimmer. Because the man up on the log deck bucked the log an inch short, every piece that comes out of that log is going to be trimmed back a couple of feet. The automatic trimmer is a piece of sawmill equipment that lends itself readily to the human element. Give any man 21 saws with finger-tip control and he will take a cut at every piece that goes through, whether it needs it or not. We are also asking the trimmerman—in an interval of perhaps three seconds—to pass judgment on a piece of lumber as to its dollar value, graded and cut in several different ways. Does he have the opportunity? I often feel that if the trimmerman were paid on the number of pieces he did *not* put a saw into, things would be better all around. Let us get our priorities in proper perspective and do first things first—that is where we can make the most money.

Bandsaw capacity

Let us briefly discuss a subject that seems of utmost importance to the industry. We should expect to know the capacity of any given bandsaw. It is sawmills we are talking about; that in itself should indicate the importance of the saw. Every piece of lumber, with few exceptions, is cut on all four sides and on both ends with a saw. We know the horsepower of the motor that drives the saw. We know the capacity of the carriage feed. We know the capacity of the dry kiln. Is it not more important that we should know the capacity of the saw itself, the number-one cutting tool in the whole industry? Many of you remember Willard Morss, who used to be in the Tacoma office of Weyerhaeuser Co. When I had the book *Feeds and Speeds** in longhand, I was looking for advice. I knew Willard was interested. I hit him at the right time; he had just completed a survey of all Weyerhaeuser's resaws. I would like to give you one or two comments from his summary.

Saw standards lacking

"I submit on a basis of facts shown that there is now no known standard or specification for saws, saw speeds or rates of feed," Willard Morss said. (That is the very thing we are talking about now.) Morss went on: "The nearest thing to a standard is to split a 12-inch piece with a 14-gauge saw at the highest possible speed." (That concerned the old 7-foot vertical linebar resaw that feeds 400 to 450 feet a minute, come hell or high water.) Morss wound up by saying: "The rate of feed must match the width of the piece being split automatically, without any . . . help from the operator . . . . It is vital . . . that these facts be established beyond question, and the sooner the better." That was Morss's response to the material I had.

Kerf savings explored

In recent years, great efforts have been made to reduce the nominal board size by reducing the kerf. In my opinion, the most logical way to do this has been largely or entirely overlooked.

Generally speaking, a problem that confronts the Pacific Northwest, both United States and Canada, is the reverse of that experienced in many other parts of the world. In the Pacific Northwest, in general, we overbite our sawtooth and gullet capacity, whereas in other areas they generally underbite. Every sawtooth has a maximum and a minimum bite. But how many consider that it has a very definite minimum bite?

Sawmill Feeds and Speeds: Band and Circular Rip Saws, by P. S. Quelch, published by Armstrong Manufacturing Co., Portland, Oregon.

For maximum efficiency, a tooth must never bite less than the clearance on one side of the saw plate. When this occurs, the chip is smaller than the opening between the saw plate and the wall of the kerf and it spills out of the gullet. This not only reduces saw clearance and causes heat but, also, owing to the grain of the wood, sawdust rarely spills evenly on both sides. It will always spill out more to one side of the gullet than to the other. (If you want to prove it, stand above your rig and look down to your shear board.) This tends to push the saw off the saw line. Some say that spillage increases gullet capacity. Well, of course it does. If you spill sawdust out, you have more room to put sawdust back. But I do not subscribe to this policy. Sawdust should be carried out of the cut and discharged at the bottom to free air.

Cutting, carrying sawdust

In cutting wood, there are two distinct operations:

1. Cutting the fiber. This is relatively simple. Just put on the standard cutting angles with two sharp corners and you have no problem cutting the wood fiber.
2. After you have turned the column of solid wood into sawdust with a tooth, you must collect the sawdust in the gullet and discharge it to free air. This is the problem—discharging the sawdust, not cutting the wood.

To avoid problems associated with discharging sawdust to free air, maximum and minimum tooth bite have to be determined. A maximum tooth bite is the plate thickness at 10 inches of saw width; for each inch in addition to the saw width, add 4 percent. With plate width at 10 inches, reduce by 4 percent for each inch less than 10 inches in width. Filers sometimes tell me that width is not that important. My answer is "then why do you throw a 16-inch saw away when it becomes 12 inches wide if width does not mean something to you?" That is your maximum bite. In shallow cuts where gullet capacity is ample, to bite more will likely tear the grain and cause the saw to lead, overbiting. In no case should the tooth bite be large enough to develop more sawdust than the gullet can carry.

Sawdust compression strains

Compressing sawdust into a gullet beyond its capacity puts an unbelievable strain on the gullet edge of the saw. This is reflected in saw performance. This is largely a problem in the Pacific Northwest. The U.S. Forest Products Laboratory at Madison, Wisconsin, has concluded that it is possible to build up pressure in the gullet between 1,600 and 2,000 pounds per square inch. The Madison lab states that up to 2,000 pounds per square inch of pressure is used to compress sawdust into an area equal to solid wood. Feeds that would develop this kind of pressure are highly unlikely, but undoubtedly great pressures are developed from feeding beyond the capacity of the saw.

A 2- or 3-inch space headsaw has about 2½ square inches of gullet area. You will remember in the days of the big fir—and there are still some left—that when you got into a deep cut, the cut was always nice and smooth at the top. But as you progressed toward the bottom, you got a vertical washboard—not a diagonal washboard. This was really the pressure building up against the wall of the kerf and literally tearing the summer wood from between the winter wood grain. This occurred in the bottom of the cut. Overpacked gullets were the reason.

Heavy feed: heavy saw

To make a saw stand up in heavy feeds, we must use heavy saws. A thin saw can never, under any conditions, cut as much as a heavy saw. If saws were never fed beyond their rated capacity, we could use thinner saws and maintain accuracy. So it does not seem unreasonable to ask for the rated capacity of the saw. We do not do that—we feed the thing until it snakes, then back off a bit.

Thin saws are not new

I am in my 70th year. As a boy, I started to file for a man who was then at my present age. I learned from him that sawmillers started off with thin saws. He told me that, if you could make a saw cut lumber at all, you could get your own price. He said he had made as much as four bucks a day (those were different days). As filers learned more and put up better saws, they fed the saws faster. So the filer, to protect himself, put on a heavier saw. But when he put on a heavier saw, operators increased the feed again. This went on progressively and for many years we have run the heaviest saws that will bend around the arc of a given-size wheel.

Now, of course, we are trying to come back again. We are on the right track. But let's not kid ourselves. Operators may feel that they cannot reduce feed. I have no argument. I am not trying to tell management how to run its mill. I am pointing out what to expect if you do certain things. *You* make the decision on what is best for your particular operation. But if you want high feeds, do not use thin saws. A reasonable reduction in feed may not be as costly as many think. The reverse can be the case.

Gullet capacity limits

It would appear that one of the principal controlling factors is the ability of the gullet to convey the sawdust out of the cut. The area of wood removed by each tooth equals the tooth bite by the depth of the cut. By the formula proposed in the book *Feeds and Speeds,* you can capably convey sawdust from an area of solid wood equal to 70 percent of the gullet. Now how did I compute that figure? Perfectly free sawdust is 2.5 times the area of solid wood. We spilled some, although we tried to prevent it. We can compress some before building up enough pressure to interfere with the saw performance. So it would be reasonable and sensible to take the mean average between absolutely free sawdust and absolutely solid wood. This shows that 70 percent of the gullet's capacity for solid wood can reasonably be used to carry sawdust. If the cut deepens, the bite should be reduced. If you are taking out longer rectangles, the tooth bite must be proportionately less or you are going to have too much sawdust. Should deep cuts be normal, it would improve cutting to widen the tooth space and consequently enlarge the gullet. There is only one way to get a larger gullet with any real amount of capacity and that is to make longer tooth space. You can deepen the gullet slightly. You do not gain very much, but you do increase the wearing line of your tooth, which costs more in saw steel, and increases tooth vibration. But it gives very little additional capacity. With less teeth and more gullet, it is possible to maintain proper tooth bite without overcrowding the capacity of the gullet.

In shallow cuts, the reverse can take place. If the depth is shallow, there will be ample gullet capacity but there may not be a sufficient number of teeth per minute to prevent overbiting the teeth. In so doing, it may overbite enough to tear grain, bend teeth and snake. If the average cuts presented this condition, it would be advisable to shorten the tooth space, giving more cutting teeth at the correct bite, but still having enough gullet capacity. Keep these all in relation to each other: tooth bite, gullet capacity, depth of cut and saw speed. You cannot change one without changing the other.

Bite limited by chew

If more feed is needed, it would seem sensible to increase the surface speed of the saw. Thus, tooth bite and gullet capacity would be in harmony. From this, we see the importance of fitting the saw to the job. Of course there are many variables in sawing timber, particularly in the log itself. Varying density should be considered. In researching this, I asked help from the Forestry Laboratory at the University of British Columbia. They were very cooperative; they did all kinds of research for me. We had hoped to find that density would be greater in the North than in the South, or the East or the West. After weeks of research, we concluded that the greatest density variation did not come about in comparisons of species or areas of growth but in one tree, from top to bottom.

I mention this to show the futility of considering density, as desirable as it may be. This is something that must be watched for by the sawyer. If he is a skilled tradesman, he knows hard timber when he sees it and he uses caution.

Average depth controls

The next variable in endeavoring to fit the saw to the job is the difference in depth of the cut. Mills either know or can find out their average log diameter. From this, you can get the average depth of cut. When this is known—and not until it is known—a saw can be used with intelligence. If we had the correct tooth bite and rate of feed for 70 percent of the cuts, it would be an improvement in most operations. Not only would it reduce saw maintenance; it would also improve quality and production.

Saws running more than 10,000 feet per minute may meet opposition. I believe that the slower a saw runs, the easier it is to maintain, provided the feed is in direct relationship. But you are not going to cut much lumber. It is easier on the saw to run it at high surface speed per minute in harmony with tooth bite and gullet capacity than to overfeed a slow saw with a relatively slow surface speed.

Before we had a chip market, we used to see refuse conveyors continually overloaded. To remedy this, the conveyor was speeded up to reduce the load at any one time by better distribution. Why not do the same for the saw? The same condition exists, but it is difficult to see.

I liken it to the man with a gravel contract who is hauling gravel out of a pit with 5-yard trucks. He is not getting enough gravel. Now he does not get the loader of his equipment to try to pound 7½ yards of gravel into his 5-yard truck and break it all; he puts on more trucks.

Do not overload your gullet. Bring more teeth into service when you want to increase the work load (Figure 10.1). Few realize the tremendous strain placed on a saw that is overfed. Possibly the filer is the only man who knows. To meet production demands, high speed may be necessary, so it seems advisable to fit the saw to the job even if it means increasing saw speed. Is it too difficult to have the feed limited to the correct tooth bite? Many other types of equipment have maximum controls. Why not saws?

Saw speed readout?

An indicator in front of the sawyer to register feed rate would be a relatively simple device. Perhaps a load meter indicating power consumption could be arranged. This would signal overloaded gullets or timber of greater density or dull saws. This equipment would greatly aid the sawyer and the filer in doing a better job.

The ultimate, of course, is a variable speed motor. I know that these motors cost money—there is a lot of opposition to them, in addition. Some say the inertia cannot be licked. I think it can. This would allow comparatively large tooth spacing to accommodate deep cuts. And the speed would be increased in shallow cuts, thereby increasing the feed by increasing the number of teeth per minute rather than increasing the bite per tooth.

The important point to note in following any of the charts in *Feeds and Speeds* (there are a number of charts with formulas showing saws with equal specifications and equal cut depth) is that the chip load per tooth remains constant throughout the various

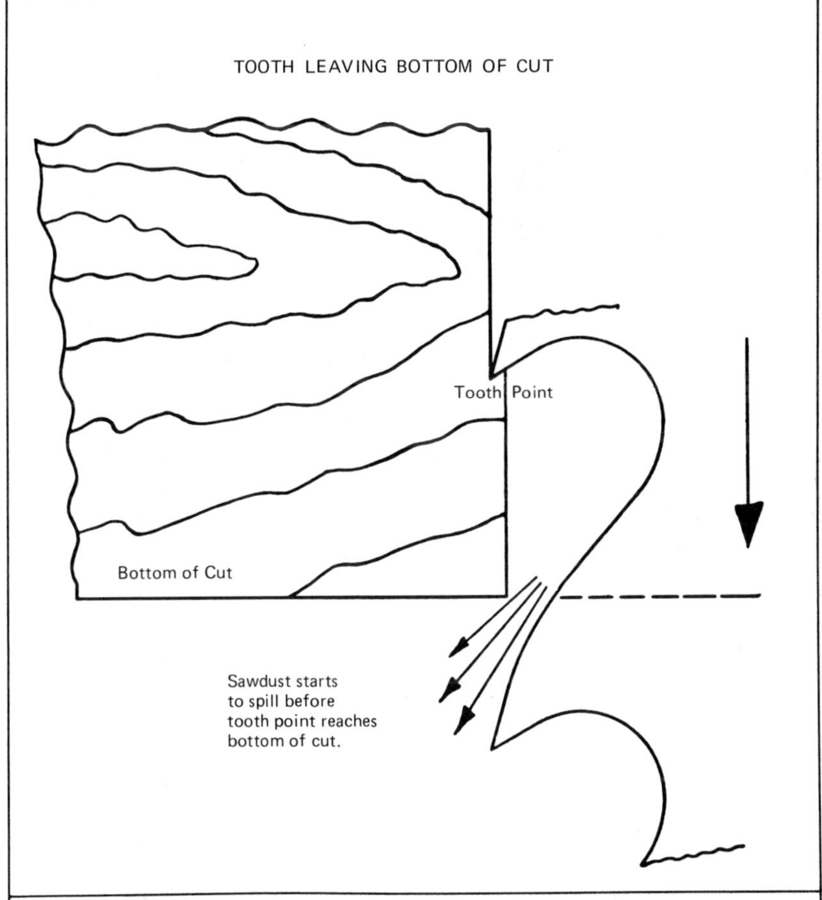

Figure 10.1. Limit the feed rate to the correct tooth bite. In this illustration, since sawdust starts to spill before the tooth point reaches the bottom of the cut, we can add 75 percent of one tooth space to the depth of the cut.

rates of feed (Figure 10.2). Rather than overload each tooth, more teeth per minute are brought into service. I would like to make it very clear that I do *not* recommend higher speeds per se. But I am saying that whatever the feed rate, the saw speed should relate. Regarding log sizes, Figure 10.3 gives a choice of saw speeds and

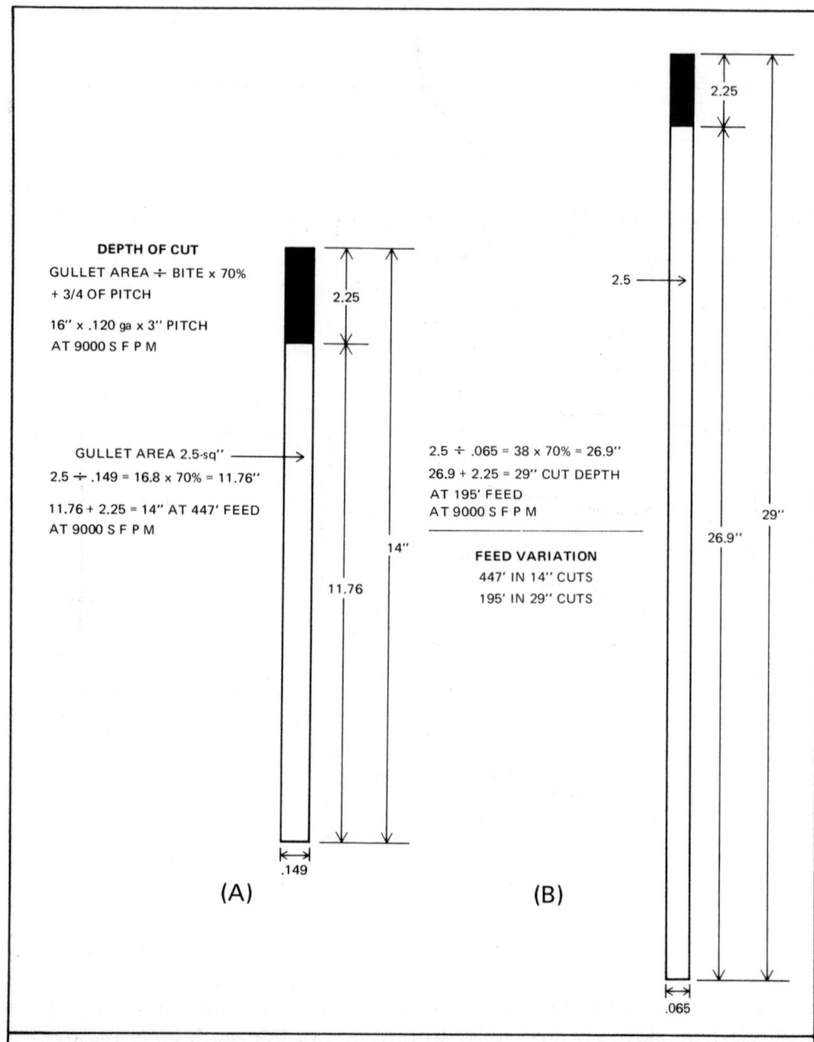

Figure 10.2. (A) Saw capacity at maximum tooth bite. (B) Saw capacity at minimum tooth bite.

tooth spacings which can be applied to a particular application. Use of this information would prevent making drastic changes after the operation has been started.

Charts balance feeds, speeds

Relationships are shown in Figure 10.4. (Chart A shows maximum tooth bite; Chart B, minimum.) The first column gives wheel diameter (7-foot); surface feet per minute is shown in the next column, then tooth spacing. Notice that tooth spacing is the same right on down to the bottom black line. Number of teeth, of course, changes with the speed. The thickness of the saw is the same in all cases; the width of the saw is also the same. Tooth bite remains the same. The total gullet area per minute changes. Maximum bite for a 14-gauge saw, 12 inches wide is 0.090; the gullet remains the same, of course. The top line of Chart A shows a machine running at 7,500 surface speed. We have a 0.090 bite,

Wheel Dia.	SFPM	Space	Number of Teeth Per Min.	Gauge	Width	Bite	Gullet Depth
8	9000	2½	43,200	.095	15	.060 .114	1 3/16
8	10,000	2½	48,000	.095	15	.060 .114	1 3/16
7	9000	1¾	62,000	.083	12	.048 .090	13/16
7	9000	2	54,000	.083	12	.048 .090	7/8

Wheel Dia.	Gullet Capacity Sq. In.	Gullet Capacity Per Min.	Kerf	% of .250	Average Cut Depth	Feed	H.P.
8	1.7	73,440	.215	86	22" 12.3"	216 410	190
8	1.7	81,600	.215	86	22" 12.3"	240 456	210
7	.812	50,344	.180	72	11.5" 7.0"	290 465	117
7	1	54,000	.180	72	16" 9.3"	216 405	120

Figure 10.3. Relationships among bandsaw sizes, speeds, tooth space, gullet capacity and power requirements.

BAND SAWS: MAXIMUM TOOTH BITE (A)

Band Wheel Dia.	SFPM	Space In Inches	Number of Teeth Per Min.	Thick-ness of Saw	Width of Saw	Bite	Gullet Depth in Inches	Gullet Area Sq. In.	Gullet Area Per Min. Sq. In.	Kerf	% of .250	Cut Depth Inches	Feed in Feet Per Min	H.P.
84"	7,500	2	45,000	.083	12"	.090	7/8	1	45,000	.180	72	9.3	337	95
84"	8,000	2	48,000	.083	12"	.090	7/8	1	48,000	.180	72	9.3	380	105
84"	8,500	2	51,000	.083	12"	.090	7/8	1	51,000	.180	72	9.3	382	110
84"	9,000	2	54,000	.083	12"	.090	7/8	1	54,000	.180	72	9.3	405	120
84"	9,500	2	57,000	.083	12"	.090	7/8	1	57,000	.180	72	9.3	428	125
84"	10,000	2	60,000	.083	12"	.090	7/8	1	60,000	.180	72	9.3	450	130
84"	10,500	2	63,000	.083	12"	.090	7/8	1	63,000	.180	72	9.3	472	135
84"	11,000	2	66,000	.083	12"	.090	7/8	1	66,000	.180	72	9.3	495	145
84"	10,000	2-1/4	53,333	.083	12"	.090	1-1/16	1.37	73.066	.180	72	12.3	400	160
84"	10,000	2-1/4	53,333	.083	14"	.096	1-1/16	1.37	73,066	.195	78	11.6	427	170

BAND SAWS: MINIMUM TOOTH BITE (B)

Band Wheel Dia.	SFPM	Space In Inches	Number of Teeth Per Min.	Thick-ness of Saw	Width of Saw	Bite	Gullet Depth in Inches	Gullet Area Sq. In.	Gullet Area Per Min. Sq. In.	Kerf	% of .250	Cut Depth Inches	Feed in Feet Per Min	H.P.
84"	7,500	2	45,000	.083	12"	.048	7/8	1	45,000	.180	72	16	180	95
84"	8,000	2	48,000	.083	12"	.048	7/8	1	48,000	.180	72	16	192	105
84"	8,500	2	51,000	.083	12"	.048	7/8	1	51,000	.180	72	16	204	110
84"	9,000	2	54,000	.083	12"	.048	7/8	1	54,000	.180	72	16	216	120
84"	9,500	2	57,000	.083	12"	.048	7/8	1	57,000	.180	72	16	228	125
84"	10,000	2	60,000	.083	12"	.048	7/8	1	60,000	.180	72	16	240	130
84"	10,500	2	63,000	.083	12"	.048	7/8	1	63,000	.180	72	16	252	135
84"	11,000	2	65,000	.083	12"	.048	7/8	1	66,000	.180	72	16	264	145
84"	10,000	2-1/4	53,333	.083	12"	.048	1-1/16	1.37	73,066	.180	72	21	213	160
84"	10,000	2-1/4	53,333	.083	14"	.056	1-1/16	1.37	73,066	.195	78	18	249	170

Figure 10.4. Maximum (A) and minimum (B) tooth bites for saws running on an 84-inch-diameter bandwheel.

which is the maximum. Depth of the cut should be 9 1/3, so call it between 9 and 10. We are feeding at 337 feet per minute. The chart gives the horsepower. Note the bottom example at 11,000 surface feet per minute. The increased surface speed has increased the number of teeth. Importantly, each tooth still is only biting on 0.090 tooth bite. We have not increased work load per tooth. But we have many more teeth. By having more teeth, we have a sawing speed of 495 feet. We have increased horsepower because there is more sawdust being removed.

Stop to imagine for one minute what would happen—and this often does happen—if the saw speed is 7,500 (as in the top line of Chart A) but the feed rate is nearly 500 feet (as in the bottom line of Chart A). I have not worked it out, but who knows what the tooth bite would be. The combination of surface speed and feed speed would probably tear the teeth out. We *do not* always keep tooth bite and gullet load relative to our rate of feed and this is what *must be done* to use a saw intelligently to gain its highest efficiency.

Minimum bites shown

Next consider Chart B in Figure 10.4. Every tooth has a minimum bite. If you bite less, you can increase the depth of the cut. Again it is the same saw in every case; now, however, our saw speeds are the same, but we have reduced the bite to the minimum; instead of 0.090, the bite is 0.048. Working that back, you will find that this is equal to the clearance on one side of the saw. Do not go below that. But by a smaller bite, we have now increased the depth of the cut to 16 inches, removing the same amount of sawdust—a taller, thinner column of sawdust, but an equal amount of sawdust.

You can work these charts out yourself. Suppose you are starting a mill or are in doubt whether you have the right saw for the job. Take out a 2-inch space. Now there is a 2¼-inch space running at 10,000—now what difference do you get? Take the 10,000 from the next to the bottom line of Chart B and run right across. Now we have a 21-inch depth of cut, but you are only feeding 213 feet a minute. Compare this to a 2-inch space at 10,000 surface feet per minute. That gives a 240 feed speed and a 16-inch cut.

You are keeping the work load of your saw relative to tooth bite and gullet capacity. You are not overloading your saw; you have exactly the same work load in all cases. If you want to put on a 2 3/4-inch space or a 1 3/4-inch space, then you see what is most suitable for the particular job in your mill.

Improvements many

We can do many things to save our raw material and, at the same time, improve lumber quality. We have many people who speak of 1/16-inch accuracy on a headrig. I have been called into many mills and when I asked, "What condition is your headrig in?" the reply was always "perfect." Yet when you run a line over the carriage you find out that it will not follow a line within 3/16 inch. I do not feel badly when people do this. We are all proud of what we do—our own little work—and we are inclined to see all the good things about its performance, and not to see the bad things about it. But consider priorities again! If we could all be more honest in our appraisals of methods and equipment, it would help. It is very difficult to get people to tell the true story. Have you ever known a man who spent $150,000 in his mill on a piece of equipment to then tell you, "Do not make the same mistake I did." Of course he won't. He will defend what he has done.

Maintenance vital

A maintenance program is important. All carriages have bed skids, wear strips on their bed skids and on their head blocks. These do not always get changed. The head blocks become loose. This, in turn, allows the dog plank to become tapered. If your dog plank becomes tapered, it will not go through your edger straight. It will lead off and get out of control of your feed rolls. The same thing happens to the gang. If you do not get a square cant going into your gang, it will lead off also. In the edger, the feed roll should be 1/4 inch above the saw line. Feed rolls should all be level and parallel to the saw arbor.

Are you as careful about tail table rolls? So many times you will see that when the piece goes out of the back end of the edger, it snipes off and requires extra trimming because of this. Why?

Down the line on your tail table, a couple of feed rolls are out of level or out of line. They take control of the piece after it has left the front feed rolls of your edger. Are we as concerned about the tail table as we should be? Lack of concern causes waste in manufacturing that, with proper maintenance, could be avoided.

It would be unfair to the maintenance department not to come to its defense. How many times has your plant come to the weekend with 21 jobs on the blackboard to do, and only men and time enough for seven? I suggest that maintenance deserves a much higher priority. This includes those things that I have mentioned—edger, trimmer, track alignment, track, carriage bed skids. The other item is the saw itself. The saw—after all these years of sawmilling—is still the most misunderstood tool in the sawmill. Being a filer, perhaps I come to the filer's defense very often. But let us be honest about it. I think the filer is the only man in the mill who honestly knows whether or not his saw is in perfect condition. There are some filers who hide behind this fact, but we do not want those fellows in the trade. The filer should be honest. But he is the only man who knows whether the saw is in top condition. This often becomes a difference of opinion between the filing room and the maintenance department. I think it has to be this way.

Tasmanian troubles

I would like to give you an example of a difference in the saw tooth load. About 15 years ago, a man in Tasmania was sent by his company to visit the United States and Canada to see what he could learn. His mill was putting in an 8-foot double cut. Before he went back to Tasmania, he bought four double-cut saws—13-gauge, 15 inches wide, 2½-inch tooth space—put up benchfitted, just as we would do it in this area. Through many years of correspondence, he told me, "I could not cut a straight board with them. But I did notice this. As the swage wore down, they improved slightly. Every time I swaged them they got worse again." What he had not told us was that he was feeding 102 feet a minute consistently, so the tooth bite was all out of relation.

We would have increased the tooth bite to get it up to where it would not spill over the side into the clearance between the saw and the wall of the saw kerf. He did not do that. We would have increased the space, or reduced the speed, or something of this nature. But he did the only thing he knew—he started to reduce the clearance. He actually got the clearance down to the point where he was running—and still runs, so he tells me—a 13-gauge saw, 15 inches wide on 0.149 clearance. We could not do it under our speeds, but this man is on 102-foot feed. I worked it backward for him to find out just what this tooth bites. The minute he got his clearance down to the size of his chip or tooth bite and it did not spill, he cut lumber. It is the same story: there is a minimum bite for any tooth.

Spruce resaw speeded

Here is another example. A little 54-inch resaw with the proverbial 7-inch-wide saw, 17-gauge, 1 3/4 tooth spaces making millions of teeth, was installed. It simply was not doing what was wanted. It was going about 160 feet a minute. They were cutting up to 12-inch spruce with it. The manager commented, "Do anything you like, but let's get something through it." I had just been waiting for someone to make that statement. So we went for 11,000, but missed it. When we got the sheaves, we were up to 11,400. According to the little bag of tricks that I worked out for him, I said that the resaw should cut 320 feet a minute in stock 10 inches deep. I had never seen one do it in my life. I thought 250 feet was pretty good performance, so I left. That is the beauty of "trouble shooting." You can go off and leave them with it! I did not hear anything for quite a while and my curiosity got the better of me. I phoned the man long distance to ask, "How is that ruddy little resaw of yours making out?" "Oh," he said, "beautiful, beautiful. We are actually cutting to depths of 12 inches with it now, not 10." About two years later, I phoned him again. "Oh, sure, it is still fine," he said. "We cut her back. We did not need 320 feet per minute. We are only feeding her 300 now—with 12-inch spruce." Now, then, if you want to work that one out, the tooth bite was correct with that speed. The filer told me he was very apprehensive about this 11,400 at first. But now he loves it and has less work than he had feeding 160 feet before. They had been overfeeding a very slow saw instead of sawing at the correct speed for the tooth bite they had.

PETER HEISER, Heiser Industrial Tooling, Seattle, Washington: Thank you, Phil, for the kind words on the Strobe saw. Does the same basic principle that you have delineated for band saws also apply to circular saws regarding too thin a blade, overloading chips, gullet capacity and all?

QUELCH: That is right. In this book there are circular saws, but I must confess I am not overly proud of the circular saw end of it. The bands I am. There have been many changes recently on circulars. The edges are an example. We used to use the old inserted bit with 13/32, 3/8, 11/32, etc. Now these are virtually all gone. We are looking at a 3/16 water-guided saw. The solid tooth is in here. But I did not include minimum bites, which I should have. These have been changing quite rapidly.

Let me warn you about this: Do not follow the book if you are using carbide. This book was written before carbide was used in sawmills. We know that it takes more horsepower with carbide simply because it is a powdered-metal tooth. We do not have normal cutting angles. I doubt that it can be quite as sharp. But as a rule of thumb, do not bite over 0.040 per tooth. Generally it is good policy to hold it down to that. I do not know whether this meets with Pete Heiser's recommendations. Pete is quite a carbide man. But this is what we have found from experience. I put a note in the new edition of *Feeds and Speeds** saying "watch for carbide." The book was not intended for that.

Sawmill Feeds and Speeds: Band and Circular Rip Saws, by P. S. Quelch, published by Armstrong Manufacturing Co., Portland, Oregon.

11

Operational experience in automated saw control

Dan Pichulo, Marketing Manager, Atmospheric
Sciences, Inc., Los Altos, California

Daniel F. Williamson, Executive Vice President
Publishers Paper Co., Oregon City, Oregon

Part 1 by Dan Pichulo

Before I tell you about operational experience with computer
saw control, I am going to tell you a story. The story, a favorite of
mine, is about a man called Arnold Flagler. As of late, our friend
Arnold Flagler has been spending much of his weekend time walking alone in the woods. One day he found himself on a very
unfamiliar path that led to a small pond. As he stood by the pond,
the water began to stir. Bubbles rose until they formed a lovely
little fountain and a voice called, "Arnold Flagler, this is the fountain of youth." Well, Mr. Flagler shrank back in fear. The fountain
leaped and soared and unbelievable music filled the air. The voice
called again. "This is the fountain of youth, Arnold Flagler.
Drink!"

"What will happen to me if I drink?" Mr. Flagler asked.

"Youth will be yours," said the fountain.

"Well, I mean, how does it work? How young will I be?"

The fountain swirled and sang as it danced higher and higher.
"Youth will be yours," it said.

"Will my family know me? What about my pension and my
cumulative profit-sharing plan?"

The voice of the fountain was fainter now. "Youth will be yours," it said.

"Listen," said Mr. Flagler desperately. "Just tell me one thing. Has anybody else tried this? Anybody I know?"

Well, the music faded away and the waters subsided. The fountain vanished and the pond was still. Mr. Flagler slowly walked home.

"What did you do in the woods today?" his wife asked that evening.

"I got lost," replied Mr. Flagler.

Future shock

Perhaps the principal reason that the Arnold Flagler syndrome exists in industry today is because of something called future shock. *Future Shock* is the name of a recent book—a best seller—on the effects of the accelerated pace of living and the rapid changes in our world on our decision-making capability and on our life style.

This is what the Sawmill Clinic is all about. It is a way to beat the Arnold Flagler syndrome.

In terms of the future shock that we see in the lumber industry, there are some important questions that really require answers. What is the market going to demand, both short and long term? What will be needed to operate profitably? How much change will I have to put up with? How quickly will these changes be needed? And probably most important: What changes will I have to make in my operation?

Key to increasing yield

When an industry development such as automatic saw control becomes an operating reality, conflicting points of view develop. This Sawmill Clinic provided an unusual opportunity to question underlying assumptions made during development. Do they still apply? Will they apply in the future?

A sawmill profits by turning logs into lumber. It is important to get every last piece of lumber, every last drop of profit, from those logs. A manager could buy and saw more logs. But there is an easier and more efficient solution: Get more lumber out of the logs on hand. When we increase recovery, not only do profits go up, but profit margins go up, too.

The key to increasing yield comes from knowing exactly where to cut each log. This requires measurement; it requires consistency of measurement and accuracy in measurement. Let us explore how demanding the measurement actually is.

The superhuman sawyer

Let us take typical log breakdown equipment with setworks. Let us say that we are operating in the range of 6 to 15 or 18 inches diameter. A typical log breakdown device may have as many as 12 sets. Of these, 40 percent require that the sawyer discern the difference between ¼ and ½ inch; 75 percent of the sets—three out of every four times he sets—require that the sawyer differentiate between ¼ and ¾ inch. The difference between sets produces an average of 13 board feet of lumber. This is the accuracy we need. Let us see how measurement is carried out in mills today.

A sawyer comes to the employment office. He says, "I am an experienced sawyer and I am looking for a job." The personnel manager says, "That is interesting. We need a sawyer, but first I will need to ask you some questions. It is very important for a sawyer to determine quickly which way the logs are moving. Also, are you sure you can keep our logs moving?"

The personnel manager continues, "Logs have different diameters on each end and everywhere between. It is important to know the difference before you set the saw. Furthermore, we want to be sure you are not confused when you look down the conveyors and see a lot of logs of different diameters at the same time. Your eyes must tell you which logs are larger, which are equal, which are smaller. We expect our sawyers to have calibrated eyeballs. You will have to determine some straight-line measurements even though you are handling round logs. And you must do this for a full eight hours."

The personnel manager continues, "Our sawfeed runs a hundred feet a minute and if you operate it for eight hours, that is 480 minutes at 100 feet a minute—48,000 feet of logs!

"And one last question: Are you going to be as productive the second four hours as you are the first four?"

Now if our prospective sawyer has survived the interview and if the personnel manager is satisfied with his answers, the sawyer will go to work. But will he do the job for which he was hired? *Can* he do the job under today's conditions, meeting today's demanding requirements?

Maybe, just maybe, we might be providing our people with yesterday's tools and expecting our people to get tomorrow's profits from them. It is important that the tools we provide be consist-

Could you pass this sawyer's test? Start at any point on any circle. Does the spiral go into or come out of the page?

ent with what we expect in production, yield and quality.

But sometimes operating pressure forces our thinking down one or two paths. For example, reducing cost is vital, but when it overshadows factors needed for effective operations some very strange decisions are made. Production can be overemphasized at the expense of yield and quality.

Electronics in the mill

What progressive, positive things can be done? For several years there have been attempts to use electronics in sawmills, particularly for measurement and control. Very early systems pointed up problems. But, in the past few years, a new generation of systems has come into use with new design concepts and new technology.

Using a quad saw, we can see how the newer automatic saw control systems are used to measure logs and control saws.

Scanning devices are mounted above the log infeed at Publishers Paper Co.'s Molalla, Oregon sawmill.

A chipping headrig with optical "dual-view" diameter and length scanners for both input accountability and set control.

Overhead scanner systems

In a quad saw arrangement, when optical technology was put together with new solid-state electronics, new types of scanners were developed. They were small and compact, could be mounted overhead and were remote from any movement of logs and equipment that could possibly interfere with or damage them. No parts of the scanning and measurement system are tied in with any operating equipment. When the log goes through the stop-and-loader and falls on the infeed conveyor, there is considerable pounding. But because the optical electronic scanner system is 8 to 10 feet overhead, vibration is eliminated. When the log is on the infeed conveyor to the quad, it is tracked and monitored constantly by the overhead scanners. The operator instantly knows length, diameter, profile and setting. With overhead scanning, there is no possibility for chips or other debris to fall on scanning components. In addition to the log movement control, the sawyer has a

Dual-view scanners take their readings at the infeed of the log breakdown equipment. Operators know length, diameter, profile and setting.

digital display that shows the small-end diameter. Set buttons indicate the optimum setting for each log.

Systems are versatile

I passed a gas station recently that displayed this sign: "Our gasoline fits in any size or shape gas tank." That sums up what can be done with the newer optical electronic systems. They fit well into just about any size mill. They seem to match equally well with any type of log breakdown equipment.

Optical electronic scanning systems can also be used in merchandisers to optimize both plywood and small and large saw logs from long log lengths. These systems can also control such equipment as beavers, skraggs, canters and bucking saws. It is difficult to think of any log breakdown equipment to which it cannot be adapted.

Is it all worth it? Do benefits outweigh cost?

Lumber up—chips down

Let us take a typical mill producing 50 million board feet of lumber. The increase in lumber is generated at the expense of chip production. In this example, credit the difference in chip price and lumber price at $80 per thousand. Fifty million board feet a year at 1 percent increase would generate an additional half million board feet of lumber at the expense of chips. At the $80 price difference, it would add $50,000 in revenue annually. In the practical world, increases of up to 10 percent are possible. With 2, 5, or 10 percent increases in yield, we can see almost fantastic returns on investment.

Let us take a typical small-log mill handling 1,500 logs a shift. Use the $80 differential between chip price and lumber price and assume that the small-log breakdown equipment is misset 1 time out of 15. At 1,500 logs a shift, we have 100 logs a shift or 200 a day misset. Assuming just 10 board feet involved per misset, we have again a half million board feet additional lumber produced— $40,000 additional revenues yearly. Project that a bit. If the misset ratio were 1 out of 10, we would see $60,000 additional revenue—or, if it were 1 out of 5, $120,000 additional. This

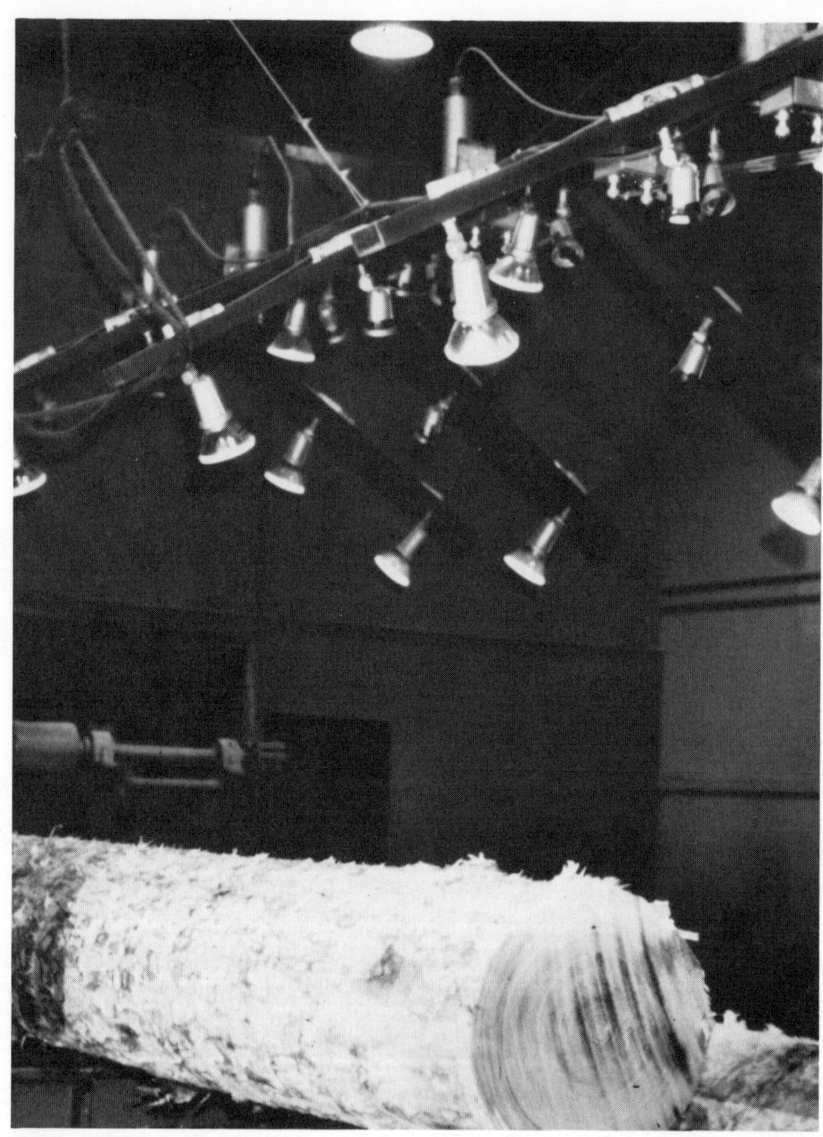

Above: The quad saw at Publishers Paper Co.'s Burney, California mill, looking across the deck with log at the stop-and-loader. Top, right: Another view looking into the saw along the infeed chain. Bottom: The control panel on the Atmospheric Sciences set control system at Publishers' Burney operation. Dual-lighted push buttons are used to improve "man-machine" interface. Unique at this particular mill is the operator's position, which enables him to look down the infeed chain directly into the quad saw. The view at top right is exactly what the operator sees from his control panel.

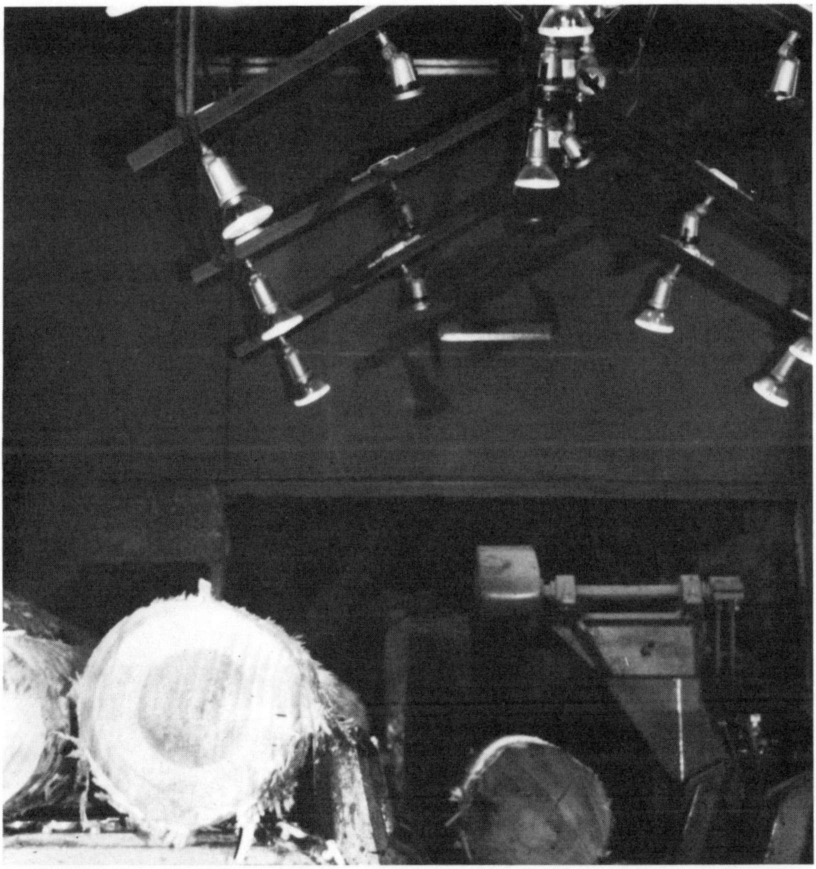

assumes that missets occur only from one set to an adjacent set. For example, if missets occurred two sets apart, then the amount of additional lumber would probably double.

The management report

Another major benefit is the means to make mill management easier and more effective. I am referring to a management report. This information becomes available because measurement has already taken place. It is now a question of assembling it. Everybody has production information, but a management report provides those very few necessary parameters to tell a mill manager exactly what is happening. Managers seem inundated with information. Trying to bring order out of it is difficult.

At the top of the report is a piece-count table. Log length is listed across the top. Diameters by one-inch increments are shown down the side of the report.

Dual-view scanners from ASI measuring logs traveling over 400 feet per minute and automatically sorting logs into six pockets based on diameters. Installation is at Louisiana Pacific's Moyie Springs, Idaho mill.

MILL MANAGEMENT REPORT

PIECE COUNT:

	SHORT	8	10	12	14	16	18	20	LONG	TOTAL
SMALL:	0	2	2	2	5	3	2	1	0	17
4:	0	1	1	4	3	2	1	0	0	12
5:	0	2	3	3	6	6	1	0	0	21
6:	0	1	4	15	8	24	5	1	1	59
7:	0	4	19	24	27	75	17	8	5	179
8:	1	7	33	54	52	134	23	8	2	314
9:	0	2	29	35	41	153	18	8	1	287
10:	0	10	22	51	34	153	32	5	0	307
11:	0	3	13	35	15	126	6	2	0	200
12:	0	1	8	15	9	70	4	2	0	109
13:	0	0	1	5	2	18	0	2	0	28
14:	0	0	1	0	2	1	0	0	0	4
LARGE:	0	0	0	1	0	0	1	0	0	2
TOTAL:	1	33	136	244	204	765	110	37	9	1539

BY DIAMETERS:

	LINEAR FOOTAGE	CUBIC VOLUME	GROSS SCRIB.	OVER-TRIM	PIECE COUNT
SMALL:	256.8	88.08	160	17.7	17
4:	169.5	50.16	115	8.3	12
5:	305.4	122.28	270	19.0	21
6:	911.8	318.68	895	37.0	59
7:	2838.5	1116.44	4400	120.9	179
8:	4850.0	2126.28	7870	220.9	314
9:	4521.8	2401.80	10390	177.8	287
10:	4810.1	3096.92	15300	199.0	307
11:	3122.0	2329.44	12220	108.4	200
12:	1718.0	1485.08	8130	55.7	109
13:	446.4	446.96	2610	11.1	28
14:	59.9	77.56	380	4.7	4
LARGE:	32.6	54.36	350	2.0	2
TOTAL:	24042.8	13714.04	63090	982.5	1539

BY LENGTHS:

	LINEAR FOOTAGE	CUBIC VOLUME	GROSS SCRIB.	OVER-TRIM	PIECE COUNT
SHORT:	7.7	2.76	10		1
8:	323.2	151.92	610	54.0	33
10:	1531.9	778.92	3530	137.5	136
12:	3154.6	1719.76	7140	148.0	244
14:	3154.0	1684.96	6130	233.3	204
16:	12799.7	7643.56	38550	301.4	765
18:	2116.1	1193.00	5070	99.3	110
20:	764.7	439.36	1780	9.0	37
LONG:	190.9	99.80	270		9
TOTAL:	24042.8	13714.04	63098	982.5	1539

Sample of the mill management's shift report. At top, the piece count table with log length listed across the top and diameters in 1-inch increments down the side. Center table gives the statistics on diameters processed. Final section of report deals with the breakdown of the shift's production by lengths. During this report's hypothetical mill shift, 13,700 cubic feet of logs were processed. Mill management could relate this to the lumber tally, converting board feet to cubic feet to determine recovery.

Let us examine the report on a hypothetical shift: The bottom line shows 1,539 logs processed, and how many of each length. Diameters processed were between 8 and 10 inches with an abundance of 9-inch diameters. The first column shows 24,000 linear feet of logs handled. This number can be related to conveyor speed. If the conveyor ran 100 feet a minute and operated for 480 minutes, we would have the potential to put 48,000 feet of logs through the saw. With 24,000 feet of logs going through, we have a saw operating at 50 percent efficiency.

In the column marked overtrim, this report tells the mill manager about the effectiveness of bucking. If a log comes through that has been bucked to 16 feet, 3 inches, it will make 16-foot lumber. Suppose it is bucked 16 feet even. Eventually, it will be cut back to 14-foot lumber. That 2 feet in the measurement would be allocated to the 14-foot column. In the hypothetical shift, logs were bucked with nearly 1,000 feet of overtrim, either on the deck or in the woods. In 14-foot logs, 233 feet of overtrim existed on 204 pieces. We had more than a foot of overtrim per piece. In the center column of the report is a breakout by diameters in 1-inch increments. This column is marked cubic volume. It shows the true cubic volume of the logs, calculated from the diameters and measured lengths.

In our hypothetical shift, we have 13,700 cubic feet of logs. We could relate this to lumber tally, converting board feet to cubic feet to determine recovery.

Report immediately available

This management report is available instantly. It could be produced at 10 o'clock, at noon, and at 4 o'clock. A manager can see the trend developing. It is a feedback report, really. If a manager makes a change in operating procedure, or adjusts equipment, he can track changes immediately to know their effects.

Stumpage prices have increased. Logs, in some cases, have been exported at a log cost greater than lumber cost in this country. The ecology scene is exerting its economic pressure. Supply is lacking, demand is going up and so are log prices. OSHA and other government programs have their effect.

Yet all we have to work with is a log. That is the one—the

only—raw material. It represents 60 to 70 percent of operating costs. Lumbermen lack options available to other industries to increase yields. In sawmilling we have two basic decisions affecting profitability. One is whether to buy more logs. The other decision is whether to get more lumber, more profit, out of those logs.

Part 2 by Dan Williamson

About 16 months ago, we at Publishers Paper Co. decided that it was necessary to reduce the burden of control on the sawyer. We needed a unit to control some saws automatically and give us better yield.

Prototype installation at Molalla

The prototype installation was made at our Molalla, Oregon sawmill, not a new mill. Optical scanners at Molalla were installed overhead. The television camera shows the operator at the control panel the discharge side so he can be sure it is clear.

The photographs accompanying this chapter illustrate most of the sequence of operations at our Molalla mill.

The minicomputer-teletype combination produces what we like to call the superintendent report rather than the management report because it is the superintendent who can actually do something about the report results—management generally just sits back and criticizes.

Reducing operator decisions

Why did we settle on electronic equipment? All of us realize that we must get the maximum from our logs. This means we have to meet the demands of the market and get the highest price for every board foot we produce. One way of achieving this is to reduce the number of decisions that have to be made by our operators.

The key decision is made with the first blade that hits the log, right at the front end of the mill, and that is where we started. We have two quads operating under control at Molalla, running from 3

A log at Publishers Paper Co.'s Molalla, Oregon sawmill moves to the quad saw headrig. Note mounting of scanning lights above to eliminate vibration.

to 5 logs a minute. When everything is clear and the feed is right, we run up to 7 logs a minute.

None of us could actually make the proper decision every time, eight hours a day. It is not humanly possible. But it is a very simple thing to do with an electronic device such as the one we have installed. The computer configuration, I am sure, could pump from 10 to 20 times as many logs through, so far as its decision-making capability is concerned.

We will be sawing to the market in the future; we are going to plug into our system the variation of market prices. Since prices vary for various woods, thicknesses and lengths, our input of this information will reflect these changes so that our programs will give us the best market price at any given time.

History of the installation

Once the decision to proceed had been made, we investigated and concluded that there were perhaps two suppliers in the business who might do the job well. We chose the Atmospheric

Sciences (ASI) group. The order was signed late in October or November of 1971. By January 1972, we had the first prototype in operation. This took about three months, which I thought excellent performance by the people at ASI and at Publishers.

We goofed on this first prototype, however, by working at the management level rather than bringing our production people into the picture. We did not bring in our sawyers or foremen; we did not involve our maintenance men to any great extent. As a result, we called a halt to the whole project. ASI went back to the drawing boards and we got moving on a new program involving the people who would be taking the action and producing the results that we all wanted.

Log, positively set, moves into the headrig.

Log moves through the saw.

Successful operation at Molalla

About a year after the original idea was conceived, we started up our first successful operation at Molalla. It has been running ever since with minimum maintenance. Also, the sawyers are very happy with it—and this is extremely important. Every once in a while when I am visiting the operation, the sawyer says, "I hardly had to take over at all today—it ran right through." He is very pleased when that happens.

The operators are pleased with the operation—they are with it; they are keen about it. Every time you visit with them, they want to talk about it.

I think we have proved that management, in developing new methods, must work with the people at the operating level who are going to be directly involved. Otherwise, the chances of success are diminished and, as in our first abortive attempt at Molalla, we are apt to end up having to start all over.

Installation cost and results

What do the installations cost? We have about $50,000 installed cost in each plant. This figure includes software development, hardware and installation.

About 95 percent of our logs are now sawn with automatic sets, and minor program changes now being instituted will increase this percentage.

You may wonder about the number of cutting patterns. At our Molalla mill, where we have from 4- to 18-inch diameters up to 24 feet long, we have 2,000 cutting patterns. At Burney, running the same diameters 8 to 20 feet long, we have 2,800 different cutting patterns.

Specific economic gains in terms of yield, product size, upgrade and the like are not yet available. We have made a number of estimates, however, and these estimates indicate that we will have 100 percent return of our investment (after tax return) within the

The prototype control panel. When operated in the semi-automatic mode, the operator simply presses the "accept" button on the joy stick to implement the measuring system's set.

first year of operation. We do not think there is any doubt about achieving that return. We will do a post-audit, probably in four to six months, to see what has actually happened.

Future plans

What else is Publishers going to do—or what are we doing now? At one mill we are adding a sorter tally to our superintendent/ sawyer/foreman report printout. We will know what comes in and what goes out. This additional information will be available on demand. The operator will have to make allowances for what is on the decks and in the mill, but he will be able to get the printout of the additional information anytime he wants it. When he is walking by, he can just push a button and out will come the additional report. The superintendent and manager will, I am sure, be reviewing this printout—particularly at the end of each shift.

Closed-circuit television gives the operator a view of the headrig's outfeed side so he can be sure it is clear.

We are starting to work now on developing a log bucking program with varying market prices to maximize return on revenues—a programming innovation mentioned earlier. We will develop a program where we will be bucking logs, which come mainly in full length, to suit current market conditions. I think most of you know that 12-foot lengths sometimes command another $20 a thousand compared to 16s and 18s. That being the case, we will be bucking the 12s to 24s and picking up that extra revenue. That is the sort of thing we are moving on next.

As to future prospects, we are considering in-process inventory control. The inventory control would go right through the mill. We think this can be tied in with our information printouts so we end up with a tidy management report.

While we do not have specific results at this time, we are confident the development program will be successful and contribute to better utilization of raw materials at all our mills.

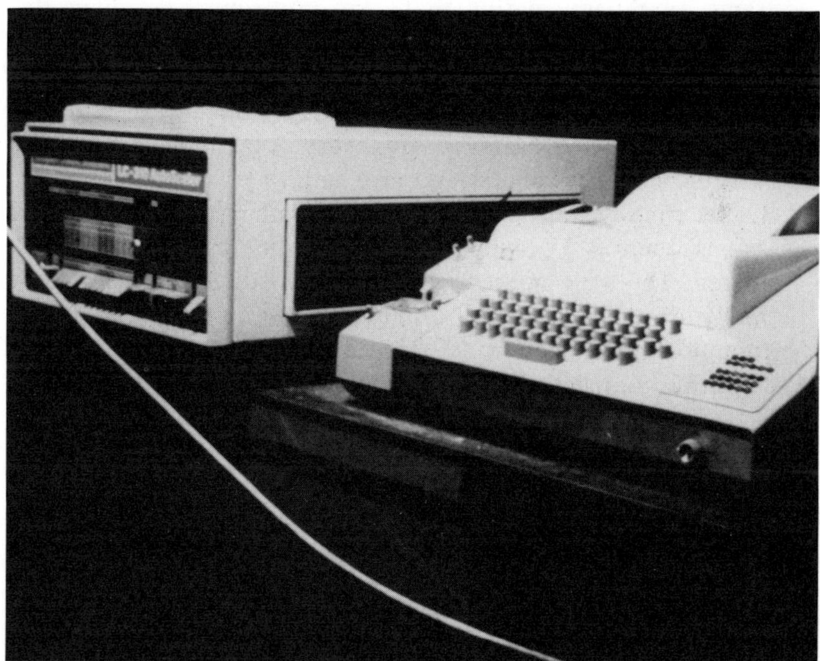

Minicomputer/teletype combination, which controls the quad saw system and prints the management report on-demand.

DISCUSSION PERIOD

MILTON MATER, Mater Engineering, Corvallis, Oregon: This is a nuts-and-bolts question. Our experiences with the optical photocell have been that a certain amount of the fine sawdust and moisture in the air in the sawmill tend to coat it, and this makes for variation in its activity. I am wondering how you overcame that with your optical systems?

WILLIAMSON: The best person to answer that is probably one of the men right in the mill. My understanding is that the units we are using have caused very few problems. We did have problems with the original prototype where, as I mentioned, we goofed. But with the primary sensors located well up above the logs, problems have been very minimal. That does not mean you do not have to clean the units once in a while, because you do.

RICHARD HARPER, Wellington Lumber Co., Warrenton, Oregon: You said you had 2,000 or so individual cutting programs. Could you define that? Are you saying you have the capability of 2,000 individual saw sets or a combination of saw sets?

WILLIAMSON: What that means is that there is a possibility of setting up cutting patterns, say, every 2/10 inch in diameter, every so much in length, and also a third parameter of taper. By the time we juggle around all the combinations and permutations, there is a possibility of 2,000 combinations to be considered. The little minicomputer looks at them instantly, then finally draws it down to either 8, 10, 12 or 16, depending upon how many specific sets are there.

BEN JONES, Longleaf Industries, Inc., Columbus, Georgia: I am like this gentleman, Arnold Flagler. I would like to see one of these things before I try it. Could we visit your plant and see these installations?

WILLIAMSON: You would be very welcome. Rather than contacting Publishers Paper directly, however, I would suggest you contact Atmospheric Sciences. I am sure they are set up to arrange a visit for you. We have no secrets in our organization, gentlemen. You are all welcome, but please make arrangements through Atmospheric Sciences as our fellows do have some lumber to produce at our Molalla sawmill.

BILL McCARTY, Potlatch Forests, Inc., San Francisco, California: In Idaho some of our timber is not quite as well shaped, or as straight and nice as the coastal timber is. Has your technology developed in terms of its programming capabilities so it can handle some of the more erratic characteristics in a log—not only taper, but some of the crooks, and so forth, that we have? Or have you at this point in time gone only to the extent of measuring just two or three key variables?

WILLIAMSON: Are you sawing those logs now? If so, I would think that the equipment would do the job. It will have to make the same kind of decisions the sawyer does. It will probably be able to make them a little more accurately. You are going to have to make some compromises—we have to make them all the time—but I would think it would do the job for you.

JIM PORTER, CAE Machinery Ltd., Vancouver, British Columbia: This is addressed to Dan Pichulo. Your company, as I understand it, now supplies the electronic equipment only. Do you intend to extend into the controls—for example, saws, the starting and stopping of conveyor systems as related to this—or are you going to stay strictly with the straight electronics so that somebody else has to get involved in the other areas of control?

PICHULO: Jim, was your question mainly "Are we going to get involved in more of the control function than the measurement function?"

PORTER: Well, often the whole thing has to work together. As a machinery manufacturer, I am interested in tying down the area of responsibility of control into one supplier's hands. I wonder whether your company has any intention of moving in this direction.

PICHULO: Our basic intention is to make sure that everyone who sees the need for automatic saw control gets a return on his investment. Our philosophy is to make sure that return on investment takes place within the first year. Whatever is required to accomplish that, we can do. We work very closely with people who are in the sorting business, for example. Where they have the technology for sorting, we can certainly interface. We are flexible enough to shift gears; if they want to take on, say, the electronic tie-in, the wiring to a particular control panel, we would be happy to have them do it. On the other hand, at Publishers' Burney, California mill, we took on the entire job and provided the control panel. We are flexible. The important thing is to get a return on investment.

MATER: This question is a little more technical. I would like to ask Dan Pichulo about maintenance philosophy. Do you provide chassis or redundant circuits, or what do you do about maintenance of this very complicated electronic equipment?

PICHULO: Regarding the maintenance of this simplified, new-generation electronic equipment, everything is on plug-in modules. Even the scanners have a little plug-in connector. It is hard for me to conceive of anybody really having a serious maintenance problem because of the plug-in replacement. On the other hand, we do provide service contracts for both preventive maintenance and on-demand maintenance, should it become necessary. We try to have a standard line of components. They are stocked on our shelves. The modules are quite small so we can always put them on a plane offering what they call counter-to-counter service. You can send one of the components anywhere in the country in about 24 hours if a plane is traveling.

WILLIAMSON: By the way, both our systems have manual over-rides that are used less than 5 percent of the time.

12

Computers in the mill environment

Dr. Don Anderson, Manager, Software Development
Black Clawson Co., Everett, Washington

Automation is certainly not a new subject. Increasing productivity through automation has a long and venerable history in the sawmill industry. I remember when I first heard the term *high-strain sawing,* it conjured up images of two men—one on top of the log and the other in the saw pit—pulling a saw back and forth.

Today mills turn out lumber by the millions of board feet—dramatic evidence of great strides forward in sawmilling. As each year passes, a little less labor goes into each finished piece. The forest industry has made strong efforts to increase both the utilization and the yield from each stem. This effort for increased recovery of high-dollar-value products will probably receive major emphasis for the next decade.

Production versus recovery

The production volume versus recovery equation will need to be reviewed in the future because we at Black-Clawson now have available accurate scaling information with which to make good management decisions. In the past, lack of accurate scaling information has been a major factor in making it impossible to determine whether a slight change in machinery improves recovery. Some mills have now reached the point where, if it were possible to increase recovery by adding a man, it would certainly pay to do so. Processing speeds, however, are such that there is little chance that a human operator can do much to improve accuracy of cutting. The answer to many of the recovery problems lies in judicious use of general-purpose digital computers coupled to an appropriate scanning system.

It is not so much a matter of replacing a man with a machine, as had been thought in the past, as giving the man a machine to carry out those functions that he himself cannot perform as effectively.

Digital computers

A general-purpose digital computer contains three major components. First is the memory component (Figure 12.1, upper left). This unit stores both the data and the instructions to execute. This stored program is the brain behind the computer.

The central processing unit (C.P.U.), the next block over, executes the actual instructions. It fetches sequences of yes and no signals from the memory and executes them in sequence to produce such esoteric things as best opening face cutting programs. But this occurs only if the computer is told exactly how to do so. This system would be no good at all if it could not speak to the outside world, so there is the necessity for interfacing this machine to readouts, displays, switch settings and sensors. These are simple devices in the external world, each providing a yes or no answer to a particular question. But when added and sensed repeatedly, these devices can provide millions of bits of information for a computer to process. With appropriate programming these samples of the outside world may be used to re-create a "picture" of that world inside the computer.

The computer is capable of executing only very simple instructions. These instructions permit adding two numbers, "and"ing them, "or"ing them, moving from one place in memory to another, subtracting, testing for zero and changing the execution sequence.

Of course, in order to read the devices outside the machine, you must be able to sense the condition of a set of test lines that connect to the outside world. All the instructions that the computer is capable of executing can usually be written on the back of a couple of 3x5 cards. However, the combination of these instructions into suitable software (the programs that actually perform the operations) enables you to create some very complex tasks.

Figure 12.2 shows how to make this machine useful. If you want a simple-minded machine to be useful at all, it has to remember everything it is told, with no memory lapses. It is desirable to have it operate very fast. The instructions mentioned above typically operate in about one microsecond. One million of them will occur within one second.

The coding of instruction sequences is the most important phase of the computer automation task and generally represents about 80 percent of the cost of the computerized control units. Programming the computers for mill control is expensive, but slight variations in its quality can change recovery from poor to excellent, and vice versa.

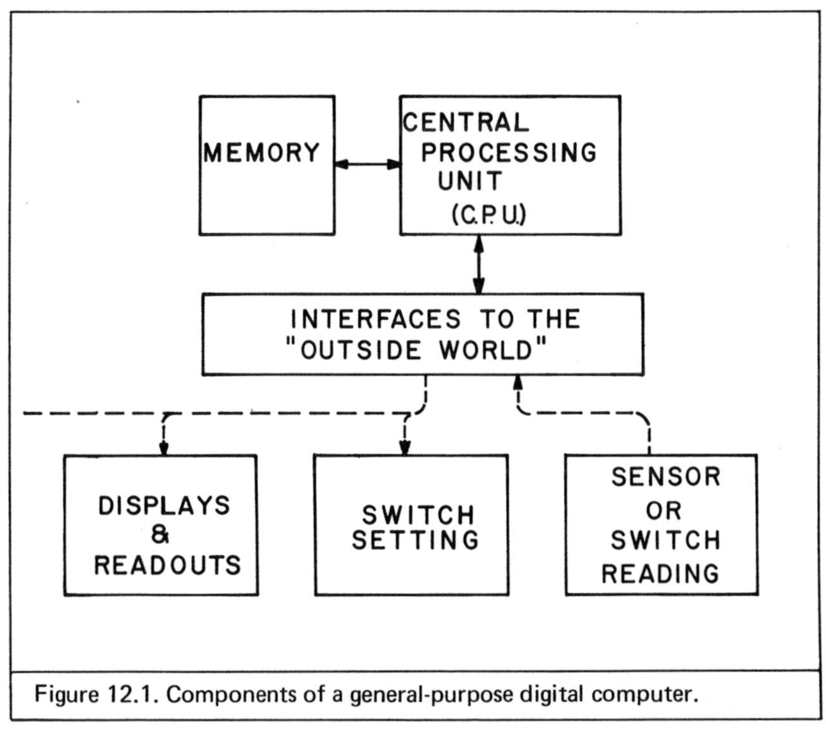

Figure 12.1. Components of a general-purpose digital computer.

Computer environment

What about the mill environment for a computer? People accustomed to data processing centers with lots of air conditioning and an antiseptic environment may wonder how a computer can survive in the usual mill environment of smoke, dust, oil fumes, water and extremes of temperature.

The modern minicomputer is quite a bit more rugged than its predecessors. But you are asking for trouble if you pour water on it, try to run it below freezing or above 130 degrees Fahrenheit. The circuits in these machines have longest life if they are kept at typical room temperature. Basically, what is comfortable for your sawyer (if he is wearing shirtsleeves) is good for the computer.

Electrically, the large current drains that are found in a mill

HOW TO MAKE A
"SIMPLE MINDED" MACHINE USEFUL

1) HAVE IT "REMEMBER" EVERYTHING IT IS TOLD.

2) HAVE IT OPERATE VERY FAST.

3) GIVE IT ACCESS TO THE OUTSIDE WORLD.

4) HAVE A SMART HUMAN TELL IT WHAT TO DO!

Figure 12.2. Four ways to make a computer useful.

may also cause trouble. All a computer requires is a little bit of 110 voltage, but if the 110 is not consistent and occasionally drops to 95 or 90, even briefly, this can cause a short interruption in your production schedule.

Stud mill use

Now what about an automated stud mill, or random length mill for that matter? There are a number of places in a mill where the computer can be used—scaling; sweep detection and bucking; headrig breakdown control; cant breakdown control; trim saw control; flow monitoring; and production records. Let us cover just a small part of one of these aspects of mill automation—the cant station. To discuss control of the cant station—either of scanning followed by gang edger or with a wane chipper between—I should probably first ask, "Why bother?" There is no point in putting a computer into the cant station unless it saves money. And it certainly does save money!

Based on studies carried out by Black-Clawson late in 1972, it appears that some two billion board feet of lumber are lost in North America every year through missetting at the cant station. Figure 12.3 illustrates this loss. Through the courtesy of a customer of ours we were permitted to go into his mill and evaluate his existing edger's performance. Cants were run through a Schurman edger. A linebar was used for setting. We found that 9 percent of the lumber potential of this station was lost—sent to the chipper. In this case, the loss amounted to an average of *2,426 board feet* per shift. Samples taken at uniformly spaced intervals over three separate days were averaged. This amounted to $255 per shift at prices prevailing then, and assumed utility or better grades would have been recovered. For 480 shifts, that would be a loss of $122,400 per year. This is not a projection. This is a measurement of what one of our customers was doing with his current set-up. We watched operators' settings to determine whether they were putting in their best efforts. Within the time limitations available to them with this production rate, they seemed to be doing an excellent job.

A good reflected light scanner and computer, combined with the appropriate setworks, should thus easily pay for itself within the first year.

Wane edges pinpointed

So much for why we should be involved in setting cants with scanners and computers. What is involved? Figure 12.4 is a very crude schematic of a cant station with a transfer scanner—feeding it through the scanner into the accelerator and into the edger. This mode of operation gives the chance to pick up the position of the edges of the wane and the edges of the cant with great accuracy— usually to the nearest 1/10 inch. Then set pins are positioned at the accelerator location to orient the cant and set it for the right saw pattern. The edgers need not have movable saws.

Figure 12.5 illustrates an idea we had for scanners. One is in operation in one of our customer's mills. This is a top-light scanner. The trapezoidal blocks on this line are schematics of cants. The light source will illuminate one wane edge at a time in the top

CENTER CANTS THRU AN EDGER

(LINE BAR SETTING)

— — — — — — —

9% SETTING ERROR RATE

2426 BD. FT. LOST / SHIFT

\$255 AVG. LOSS / SHIFT

\$122,400 AVG. LOSS / YR. (480 SHIFTS)

Figure 12.3. Average annual loss in lumber and revenue in North America through missetting at the cant station.

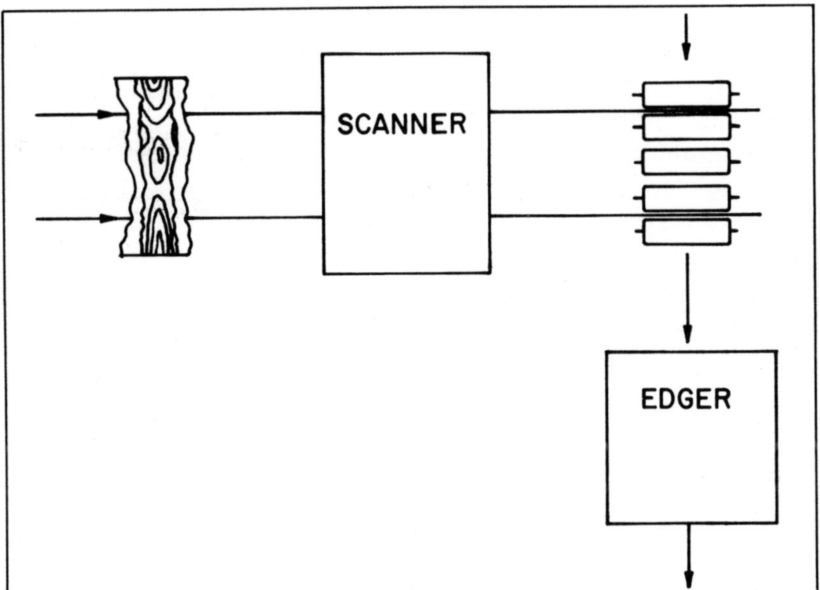

Figure 12.4. Schematic of a cant station with a transfer scanner. At the accelerator set pins are positioned to orient the cant for saw pattern.

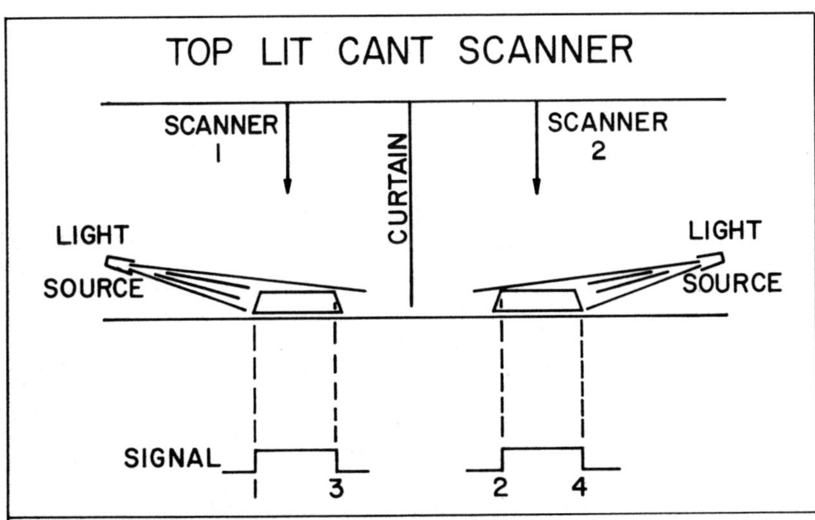

Figure 12.5. Design for a top-lit cant scanner. The light source will illuminate one wane edge at a time in the top surface of the cant.

surface of the cant. Then a scanner on that edge will pick up a signal, lit or unlit. From that signal, the first scanner tells us about the first edge, which is a cant edge, and the third edge, which is a wane edge. Further on, a second scanner with light coming from the opposite direction can pick up the second edge, which is the wane edge, and the fourth edge, which is the long edge of the cant. This information is fed into the computer to calculate the cutting program. Physically, about 4-inch spacing along the cant provides very good data. Positioning of the edges will normally be to 1/10 inch.

Figure 12.6 shows another design for a scanner. In this case, we orient the light only at the wane edge, keeping the two surfaces as dark as possible. One advantage is the possibility of using only one

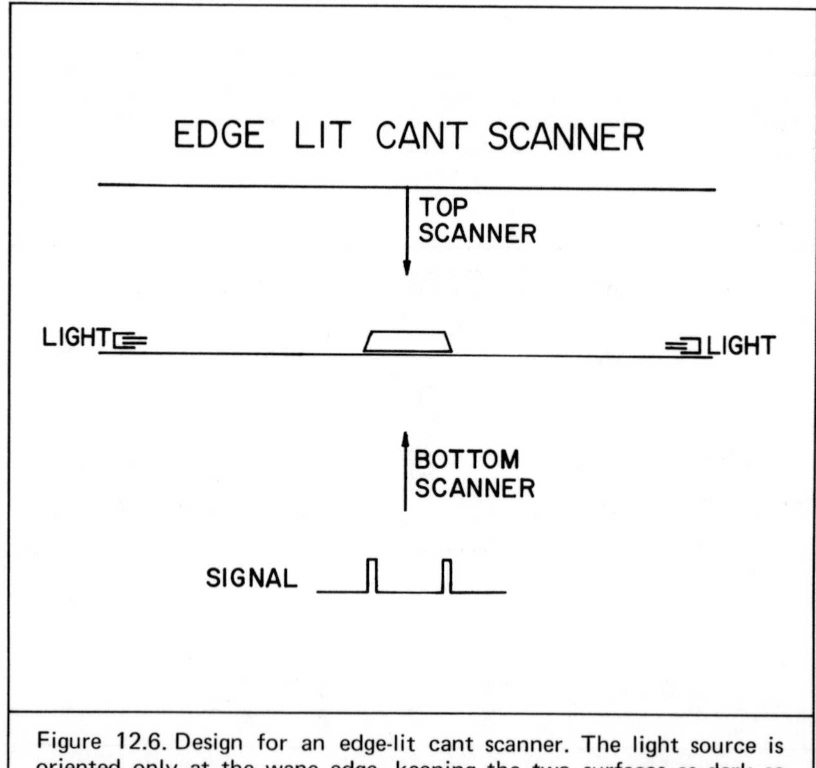

Figure 12.6. Design for an edge-lit cant scanner. The light source is oriented only at the wane edge, keeping the two surfaces as dark as possible. Further improvements are being made on this prototype.

scanner. In this type of application, however, I think it might be helpful to have a double scanner. This would avoid the necessity of turning the cant. Cants could be chipped either right side up or upside down—it would not make any difference. The light illuminating the edge will produce a signal that is the width of the wane.

This scanner is not an installed version. It is just an engineering prototype, and we are making further improvements in this model. Problems that can be encountered in this type of operation relate primarily to the accuracy with which the cant can be moved past the scanner. In using a moving log scanner, if the cant is not referenced to a set of pins as it moves past the scanner, the scanner data will be offset by the amount that the cant is not against the pins.

When the cant reaches the accelerator, it has to hit a complementary set of pins and be set. A smooth flow of cants with the chain moving fairly continuously is highly desirable in this area.

Many mills with modest investment can refit to take advantage of the savings inherent in this type of operation; whereas to change a headrig or install scanning in that area would be a substantial problem.

Several small computers seen

A computer decision that will come up for most sawmillers within the next 10 years is whether to choose a computer that handles all the functions in the mill or to have a separate computer to handle each major station. I think that the best bet in the long run is to use small computers at each station. The central processing unit cost for these small computers is becoming a minor part of the cost of the total system. The mill need not be rebuilt to install a small computer system. It is easier to pin down malfunctions if you have a computer assigned to each task. Malfunctions tend to accumulate as you give a computer more tasks.

Programming is the most costly portion of the operation. A single centralized computer makes the programming more difficult, hence, more costly. Smaller computers put fewer eggs in one basket. If you choose one vendor to set up a system for your entire mill, you *are* putting a lot of eggs in one basket. By taking it a piece at a time, you have the opportunity to make a modest investment and to see how good the vendor is. The smaller systems also provide an opportunity to familiarize your employees, one station at a time, with computer operation. Automation requires psychological adjustment on the part of the human work force. We have had many cases where the operator felt he had to override the computer at every opportunity. But what he was really doing was destroying a lot of good wood. It takes a little adjustment period to convince the operator that the machine, in most cases, is improving his cutting. And that, of course, is the objective.

13

Recovery at the resaw: band vs. gang

Jack Gates, President, 3-G Lumber Co.
Philomath, Oregon

Sawmill operators are traditionally known for their rugged, individualistic views on how to convert logs into wood products. Sawmill operations are almost as varied as people. I am sure some, or all of you, will disagree with at least some of my views. This is fine, because if all sawmills and sawmill operators were the same, life would become very dull and routine.

In this paper, I am going to discuss some of the advantages of a resaw over a cant gangsaw for a sawmill such as 3-G Lumber Co. Our logs consist of large second-growth timber with coarse grain and correspondingly large knots. We believe that as time passes more and more mills will be cutting this type of log.

At the headrig, we saw all logs of 18 or more inches in diameter to a 12x12 cant with the pith as nearly centered in this 12x12 as possible. Logs under 18 inches in diameter are sawed to the largest square possible with the pith centered. The 12x12 size is reducible to whatever dimension lumber size is most advantageous for the lumber market at the time.

Knot size declines

Knots within a log are cone shaped; no matter how large they are on the surface of a cant, they diminish to zero diameter at the pith. This brings us to the first advantage of a resaw for reducing cants to dimension lumber.

As we start sawing a 12x12 into lumber, the first cuts will produce 2x12s. These first cuts are always taken from the side of the cant with the largest knots. A Number 2 2x12 will allow a 3¾- to 4¾-inch knot, depending on the location of the knot. Each successive cut is taken from the side of the cant that has the largest knots.

As each succeeding board is removed from the cant, the width of the next board decreases, which also decreases the size of knot allowable in a given lumber grade. To compensate for this, each board is closer to the pith and the knot is correspondingly smaller. This procedure is continued until the original 12x12 becomes a 4x4 with the pith still in the center. At this point, any face of the 4x4 is only 2 inches from the pith, and at 2 inches from the pith, even the largest of knots has diminished to mediocre proportions. The 4x4 can then be cut into two 2x4s.

With a gangsaw, however, all cuts are made at one time, so it is not possible to rotate the cant to align the width of board with the size of the knot as it diminishes in size.

The first advantage of the resaw is the ability to saw for grade by turning the cant to align the width of the board with the size of the knot.

Photo courtesy FOREST INDUSTRIES magazine

With a gangsaw, all cuts are made at one time. Therefore, it is not possible to rotate the cant to align the width of board with the size of the knot as it diminishes in size.   Photo courtesy FOREST INDUSTRIES magazine

Spike knots less

The second advantage of a resaw, if used as described, is elimination of spike knots and narrow face knots. Since knots affect the grade by their displacement volume, it takes a much smaller edge or spike knot to reduce the grade than a knot in the wide face. For example, a ¾-inch knot in the narrow face of dimension lumber would displace 50 percent of the cross-sectional area but a ¾-inch knot through the wide face would have very little effect.

Resaws can be used to advantage to recover the maximum of usable wood on cants with defects such as decay, shake or split. If a defect appears on a surface of a cant after it has been partially sawn, the sawing pattern may be changed at any time to recover the maximum amount of usable lumber.

Dimensions versatile

Resaws are extremely versatile in reducing cants to any desired dimension size. A 12x12 cant, for example, may be reduced to any size from 2x4s to 2x12s or to any combination of sizes to match the best market at the time or to fill order requirements. And, of course, this reduction to smaller sizes is accomplished without loss in board foot volume. Rejected pieces from orders for beams, stringers, posts or timbers may be reclaimed easily by re-sawing to dimension lumber.

Offbearing steadier

Since a resaw produces only one or two pieces per sawing line, the outflow of lumber behind a resaw is usually simpler to handle through whatever transfer equipment is required to move the lumber to the next station. The outflow from a gangsaw is in multiple pieces and usually requires some means of separation in the flow system.

Reduced planing losses

The last item I would like to mention in comparing gangsaws to resaws is one that is becoming much more important as sawing tolerances and planing allowances are reduced to a bare minimum. The resaw produces a smoother surface on the lumber than a gangsaw. This is because the gangsaw becomes momentarily motionless at the top of the stroke and begins its cutting cycle at a low speed, causing more of a tearing action in the wood. A band-saw, on the other hand, rotates continually in one direction and the teeth strike the wood at a higher rate of speed. This causes a shearing action and produces a smoother surface. This smoother surface greatly reduces loss at the planer.

14

Selecting, sawing & selling specialty items

Ray Swanson, President, Swanson Bros. Lumber
Co., Inc., Noti, Oregon

In years gone by, many mills produced what today we would call specialty products—long timbers, stringers and fractional sizes as well as various ceiling and siding patterns and millwork. Today so many mills emphasize "recovery" in terms of scant-sawn, precision-standardized sizes that almost anything deviating from those sizes is, to a degree, a specialty item.

It is perhaps an overstatement to say that most mills are concentrating on the recovery or overrun aspect of manufacture, but it is true that every mill, to survive, must apply the maximum-quantitative-recovery principle to at least a part of its production.

Division, not addition

The manufacture of lumber is essentially a separation process, in contrast to most other material manufacturing processes. While most other products obtain their maximum structural and economic value by combining various elements advantageously, lumber, like precious metals and valuable jewels, reaches its highest value by dividing its combination of characteristics into a variety of sizes, shapes and patterns. As with fine gems, once the separation has occurred, there is no second chance to recombine the elements and try again. The highest value is achieved by the judgment and workmanship of the artisan.

It seems to me that specialty lumber products today include anything that does not fit into the modular minimum-size pattern for ordinary dimension—boards and timbers. Probably most specialty products are cut to a fractional size.

Price the equalizer

I believe it fair to say that the usual practice in cutting fractional material is to sell it on a size that approximates more closely the finished size than does standard-dimension lumber. Thus the recovery factor is lower and the difference must necessarily be overcome in the price. I believe it unfair to say that a mill that does not obtain a recovery factor of 200 percent is wasting trees because it does not cut everything into dimension.

Critical skills

If one factor is more critical than any other in specialty item

Bucking in the woods can help channel the product toward its maximum economic and structural value.   Photo courtesy FOREST INDUSTRIES magazine

manufacture, it is the attitude and the proficiency of the people who are engaged in the process. From faller's saw to delivery of the finished product, more attention is required to develop and maintain the quality of the material. In standard-dimension items, if the piece is slightly mismanufactured or off grade, it can be placed in a lower grade. But the odds are that a fractional size has the penalty of loss in scale as well as grade. A working knowledge of end use as well as the printed grade rules is most helpful for each production person whose decisions are involved in establishing the quality of the material.

Selecting standing timber

In terms of a specialty premium—whether the main emphasis is on quantitative recovery or on qualitative recovery—the timber faller can make a difference by his decision in regard to the falling of the tree. In the production of specialty items, the bucker can be even more helpful in channeling the product toward its maximum economic and structural value.

Many times, the decision made in the woods vitally affects the final destination. The judgment can be altered by observation of the age of the timber stand as determined by the number of annual rings in relation to the diameter, by the color of the bark, by surface indicators, even by the sound produced when the log is struck by an ax or hammer.

Separate the butt log

If the choice could be made on qualitative economics alone, I would favor cutting the butt log for a short peeler block or stud material. However, the logistics of log handling make this approach somewhat impractical. We have followed this course on occasion, when circumstances permitted, and the results were beautiful. The second cut on the log has straighter grain, less severe taper, more freedom from pitch and stump rot and more flexibility for grouping the natural characteristics of the log into a variety of higher-value products. There is nothing new about this concept. It is the same practice used in the early days of the springboard and the high stump, except for the utilization.

In any case, the bucker can have a substantial impact on the structural and economic value of the product of the log by his knowledge of the best alternative for further manufacture.

Decisions at mill

When the logs arrive at the mill, the second selection for length begins with debarking and bucking. Here orders are written with specifications for various products and the operator makes a judgment on the best combination of lengths to fit specific orders. His decision is guided by the variety of orders that happens to be on the board at a particular time.

A high percentage of purchased logs come in peeler lengths of 17, 26 and 34 feet, so there is some difficulty in handling specifications that involve 20 and 22 feet. For that reason it is not uncommon, where we have the choice, to cut more heavily to these lengths (20 and 22 feet) in our own logging operation.

In the mill, logs are bucked for the best combination of lengths to fit specific orders.   Photo courtesy FOREST INDUSTRIES magazine

Failure to make the right cutting decision at the headrig station can be extremely costly. In addition to the monetary value lost with mismanufactured lumber, loss in scale is inherent in remanufacturing to another size.

Photo courtesy FOREST INDUSTRIES magazine

Sawyer's job

The sawyer's station at the headrig is, without a doubt, the position at which the most vital decisions are made. There is a wealth of evidence on which to make these decisions, but once made there is little latitude for subsequent operations to change them without some penalty, which can be extremely costly.

As an example, let us say that the sawyer has failed to use the right amount of taper set for a log that was barely large enough to make some 4½x12 vertical-grain shipdecking. The resulting cant would have an uneven face width resulting in fewer pieces making the grade. Or worse, the cut would be too deep on one end and would have defects on the reverse side that could lower the monetary sales value of the material by 50 percent or more, plus the loss in scale to remanufacture to another size.

A second consideration is wane or lack of wood at the corner and ends of these premium pieces. Almost without exception, premium grades of clear allow no wane, so that inaccurate taper adjustment can make excess trimming and edging necessary; if the

At the trimmer, length is already confirmed unless there is some particularly objectionable characteristic that needs to be removed or separated by cutting the stock into segments.   Photo courtesy FOREST INDUSTRIES magazine

order is a one-length specification, wane can force remanufacture of the piece to another size.

A third consequence is the additional difficulty in handling a tapered piece that impedes the smooth and rapid flow pattern through the mill. On logs that have substantial surface clear, we use the taper all the way around the log and take the wedge out of the center. That is but one facet of the sawyer's decision-making responsibility; others include calculating the various size combinations that can come from a single log and checking the accuracy of the lumber to detect variation in sawing for minimum tolerances.

Rear of the mill

The farther along the production cycle moves, the fewer variables there are to affect the necessary decisions. At the edger, there are certain dimensional limits already established. The modular size and the length of the cant give some indication of the final finished size to be obtained. If the cant is 4 inches thick, the edgerman knows that it may be suitable for 4-inch width vertical grain or some module of 4 inches for flat grain—this could be 4/4, 5/4 or 8/4. At any rate, the decision is less flexible. At the trimmer, the length is already confirmed unless there is some particularly objectionable characteristic that needs to be removed or separated by cutting the stock into segments. The same thing is true at the resaw, although the resawyer can "steal" a higher grade piece from a cant now and then.

Vertical grain

As a general rule, we try for maximum conversion of clear into vertical grain cutting and for flat grain in the structural and commons. One other practice that is encouraged at the resaw is to make the heart centers into as small a piece as the knot size will allow. We feel that this limits the number of spike knots and makes the material less susceptible to warping and checking, whether it is dried in the kiln or is shipped green to some warm climate.

Drying factors

The drying process requires considerable skill and attention, especially where there are small lots of the heavier sizes such as 3- and 4-inch sizes. As anyone who has had experience with these items knows, it takes a considerably milder schedule for the heavier material if excessive kiln degrade is to be avoided. Where sizes are mixed in the kiln, the smaller size may come out too dry, creating problems.

Basically we feel there are at least three things that are critical to superior drying performance in addition to those mechanically controlled by the kiln equipment. These are:

1. Sticking and load supports to keep the lumber straight while in the kiln.
2. Maximum sorting for length to make the loads solid.
3. Adequate baffling in the kiln—as much as possible with solid loads—and the use of artificial barriers where necessary.

We use what we feel to be exceptionally mild drying schedules, since there is generally adequate kiln space and no shortage of hogged fuel to affect the cost. There appears to be a minimum of falldown in grade from kiln-generated defects and the extra time, we think, makes good material even better.

Handling care

Specialty items require extra care in handling, too. It takes shed space for protection from the elements as well as antistain treatment for clears that may remain tight piled for a period before drying. Of course, all material on the "R" list destined for export must be treated. During summer weather, most specialty items also need end seal treatment to prevent—or at least reduce—checking from low humidity and/or direct sunlight. It is obvious that considerable extra cost is inevitable and must be taken into account in the sale price.

Communication—the key element

In one sense we could say that we play both ends against the middle in this economic game. We must be able to buy the raw material to sell and we must sell advantageously what we are able to buy. Either end of the scale starts the action.

The key element in the action program that makes possible the whole cycle of specialty item production is controlled continuity of the whole system. When an inquiry is made for a specialty item, the lines of communication and the working knowledge of the participants come into play. Let us take an example of a recent experience with some shipdecking.

The order was taken on the verbal assurance of a high-grade log purchase and sale commitment that easily would have supplied the necessary raw material, had it been fulfilled. However, for a variety of reasons—among them the volatile price gyrations in log values and the relationship to price control regulation—the log supply from that source did not materialize.

It was necessary literally to go into the woods and select trees which were large enough and good enough, hire a logger with an S-J-4 yarder and loader, bring a set of fallers to the scene and selectively log and almost escort each piece personally through the mill. Fortunately this is the extreme and not the usual circumstance, but it does illustrate the detailed intensive effort necessary to get the job done.

Marketing material

Yes, selling is an indispensable and inseparable element of the whole process of getting the product to market, as well as buying the logs. It has been said that "the craft of the merchant is to bring a thing from where it abounds to where it is costly." Anyone who has been in this business as long as I have is well aware that there has been dramatic change in this philosophy in recent times.

The kind of timber necessary in manufacture of many specialty items, as we have known them, is not unlimited. (It was considered to be so at the beginning of this century.) While it is possible to accelerate the production of fiber per se, it is unlikely that we can afford to restrict timber growth to produce annual rings at the rate of 10 to 20 per inch. Therefore, this kind of specialty item will continue to be available only in limited quantities.

The key element in specialty item production is controlled continuity of the whole system. This photograph shows a load of kiln-dried lumber, destined for use as bleacher seats, entering a vacuum chamber to soak up a preservative solution. Photo courtesy OLINKRAFT-OLIN CORPORATION

Attrition of uses

We are now at the stage where we must bring a costly product to the marketplace where it is even more costly. In my judgment attrition of markets for specialty items made from deep old-growth timber is occurring at a more rapid rate than that at which new markets are developing. Some of this is due to plant closings caused by environmental restrictions, some for economic reasons and, in some cases, because substitute materials have taken over. The current chaotic market may speed up the trend.

Many of the accounts we serve—and they are almost all handled through wholesale channels—are customers of long standing. One of them uses a substantial amount of 6/4 lumber for construction of a shrimp boat fleet. I am advised that the supply situation is critical and that there is danger of having to curtail or shut down operations because of the lumber shortage. It is an operation that has been in business for many years. These boats average about 70 feet long. The one thousandth boat came off the line 10 years ago. If you total that, you come up with 13¼ miles of boats end to end.

It is a substantial market that the specialty lumber industry serves and hopefully it will survive the present situation. Another customer is looking toward finger-jointed material to fill his needs, for the same reason. If he is successful, it will make for better utilization and more flexibility in the marketplace.

Pattern-change problems

Phase II worked a severe hardship on both the manufacturers and users of lumber specialty items. It generally has been true that price patterns and trade channels have been relatively stable for these kinds of materials. This is not true today. It is difficult and sometimes infeasible to change from one cutting pattern to another for the short-term, one-shot item, particularly if it is dry and is to be shipped as an exact quantity. The restrictions of Phase II—for the mill with a historical price structure and the utter impossibility of allocating cost increases precisely to a given product from the same log— create a prohibitive situation.

At the same time, a mill without the historical restriction could class the lumber as a new product and could charge the "market price" by whatever definition it could claim valid at the time. As a result, some mills were rumored to be changing cutting practices and product lines in self defense. It is difficult to predict what will happen during Phase III and subsequent "phases."

'Sail we must'

There is no doubt that we live and work in a rapidly changing world. This situation brings to mind a bit of philosophy quoted by Oliver Wendell Holmes: "I find the great thing in this world is not so much where we stand as in which direction we are moving. We must sail—sometimes with the wind and sometimes against it—but sail we must, and not drift nor lie at anchor."

So change we will when necessary. Until we see a better way, we will stay with the vagaries of the market as best we can. The selection, sawing, and selling of specialty items is a demanding business, but the forest industry has always been that way.

15

Trimmer control & lumber tally systems

Edward N. Rickford, President, & Gene Neely, Vice President in Charge of Process Control, North American Controls Inc., Portland, Oregon

Part 1 by Edward N. Rickford

Although North American Controls is a new name to most people in the timber industry, its principals have all had long and extensive experience with industry problems. Gene Neely has worked closely with manufacturers of sawmill and plywood machinery in the design and implementation of process control and tallying systems that restrict operator decision-making to qualitative judgments only. All quantitative measurements are electronically executed and stored in minicomputers. Our electronics engineers have had considerable experience in the design and manufacture of electro-optical scanners. As a consultant I have been recommending the need to increase recovery in sawmills through technological advances in new high-precision machinery.

Developments in high-strain, thin kerf bandmills and high-precision, thin kerf edgers make higher recovery possible by sawing to scant dimensional sizes employing thin kerf saws. Computerized log sawing techniques make it possible to predetermine in theory best sawing sequences by individual log sizes down to 1/10-inch diameter increments.

Technological developments traced

Historically, the basic philosophy in wood conversion involved the liquidation of a dollar investment in raw material as quickly as possible into profit, with the emphasis on production rather than recovery. Over the past five years increasing raw material costs have awakened the sawmilling industry to the need for improving recovery through the introduction of high-precision, thin kerf bandmills and edgers, producing accurate lumber to close dimensional tolerances with smooth sawn surfaces.

Concurrent with the advances in sawmill machinery technology, we now have the ability to theoretically predetermine the optimum sawing sequence for each individual log size, based on computer programs such as Hiram Hallock's best opening face and H. C. Mason's sawing selection program.

The next step is implementation of process control and information systems using electronic sensing devices and minicomputers to control and monitor operations against theoretical optimums. Wood fiber savings from implementation of these technological advances form the basis for justifying major capital investments in new sawmills.

Automation through process control

Our principal objective at North American Controls is to develop and implement process control systems in sawmills to automate setworks positioning at individual machine centers against optimized sawing and cutting solutions. In theory we can predetermine the optimum log sawing solutions; in theory we can determine the precise trim length for lumber. In practice the operator is incapable of eye-balling log dimensions to the nearest 1/10 inch and storing in his head the thousands of alternative sawing sequences designed to maximize dollar yield from each individual log size. In practice the operator is incapable of determining precise trim length at 50 pieces per minute within the grade rule restrictions on wane allowance. Technology is available to do this and we at North American Controls are able to install these systems. Our ultimate goal is to develop real time information systems that allow production managers to measure operating efficiency by comparing theoretical log-to-lumber recovery factors with actual results for any log size distribution.

Overrun versus recovery factor

The sawmilling industry has traditionally used overrun percentage as a measure of operating efficiency—that is, the ratio of board foot lumber output to board foot log volume input, Scribner scale. Even in operations where overrun is computed at the end of each shift, the factor is generally meaningless, particularly where there is a high proportion of small-diameter logs in the 4-, 5-, and 6-inch classes, all having the same volume, according to the Scribner table.

There are few instances where daily computation of overrun is used as a measure of operating efficiency, in the sense that the production manager takes immediate steps to investigate and rectify the reasons for a drop in overrun—whether it be due to personnel inefficiency, poor sawing practices or serious wood fiber losses at the edgers and trimmer.

As log diameters continue to decrease, the sawmilling industry will turn more and more to the use of solid cubic formulae as a means of computing log volume, and recovery factor to measure operating efficiency.

We at North American Controls provide a system that keeps a continuous record of log input, collected at the headrig, and lumber output at the green trimmer. At the end of each shift the production manager is supplied with a single number, the recovery factor—that is, the actual board feet of lumber output produced from a known log diameter distribution and cubic volume input of logs. This number will then be used to compare the actual recovery factor with the theoretical recovery factor developed from computerized log sawing techniques, which determine the optimum product mix from individual log sizes.

Operating efficiency

Operating efficiency is simply the ratio of theoretical recovery to actual recovery over the operating shift. The closer this ratio approaches 1, the more efficient the operation. For example, over a log diameter range of 5 to 20 inches, averaging say 11½ inches, the theoretical recovery factor may be as high as 11 board feet lumber output for every 1 cubic foot input of logs. Many good mill operations recover between 7.5 and 8 board feet per 1 cubic foot log input. Industry average drops as low as 6 or 6.3 board feet per 1 cubic foot log input. Improvement in operating efficiency can be measured by management's ability to increase the actual recovery factor as close as possible to the theoretical optimum, within the limitations of log quality. Using the T/A recovery factor ratio provides management with a quantitative tool for measuring potential improvement in their mills.

Improvement in recovery factor may take several forms:

1. Training personnel to become more competent in operating fairly sophisticated systems.
2. Improving precision at individual machine centers to permit sawing scant to closer dimensional tolerances.
3. Employing thin kerf saws on high-precision machinery to increase recovery.
4. Revising sawing programs around valid predetermined optimum sawing solutions.

Information system

We are limited at this sawmill clinic to a discussion of trimmer control and lumber tallying. However, the total information system North American Controls offers the industry requires collection of log data at the headrig, in order to monitor continuously the operating efficiency of the mill.

There are many competing optical scanners on the market that determine log diameter, taper and length. These scanners are interfaced to minicomputers which keep continuous tallies of log input during the operating shift. Whether or not these systems are designed to function around predetermined optimum log sawing sequences, the log data accumulated at the headrigs form the basis for relating lumber output at the green trimmer to log input at the headrig.

Figure 15.1 summarizes the total information system offered by North American Controls. It illustrates the sequence of operations

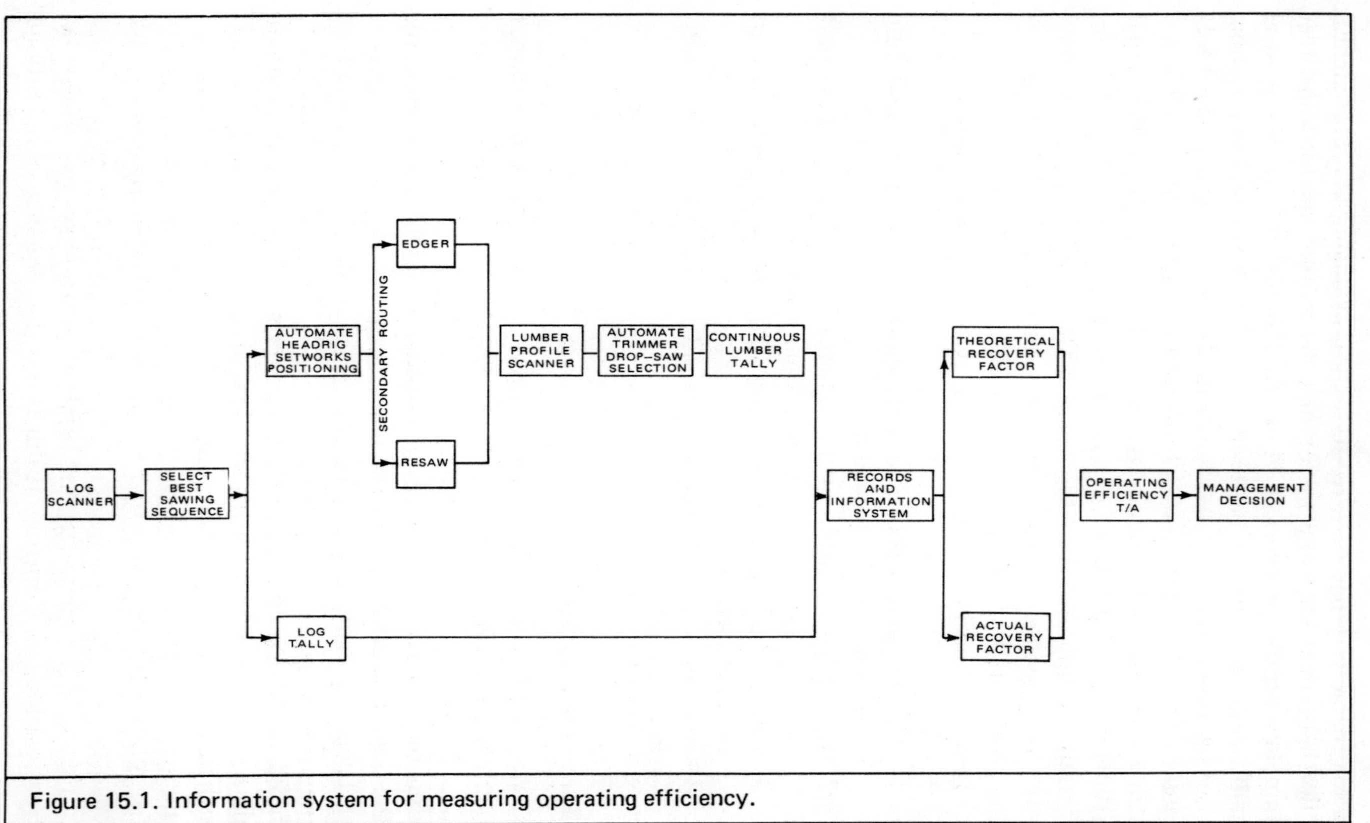

Figure 15.1. Information system for measuring operating efficiency.

from the point of scanning bucked logs at the headrig for collection of log data, to the point of tallying lumber output at the green trimmer, for computing both actual and theoretical recovery factors.

The illustration in Figure 15.1 assumes storage of predetermined optimum sawing sequences in a minicomputer which forms the basis for actuating setworks positioning at the headrig. A continuous log tally record is kept of diameter, length and cubic volume. From the headrig lumber is routed to secondary breakdown machines, such as edgers and resaws, before arriving at the trimmer. It is at this point that North American Controls can offer one of two systems:

1. A trimmer control system, designed to automate drop saw selection at the trimmer by first profile scanning the pieces travelling transversely on the infeed table, and determining the precise trim length in accordance with wane allowances.
2. A lumber tallying system which accumulates length, width and thickness of all lumber exiting from the trimmer.

Both systems incorporate minicomputers and teletypes for accumulating output data and sorting by product size and length and printing out tally reports.

Determination of the actual recovery factor is a simple division of board feet lumber output by cubic feet log input. Calculating the theoretical recovery factor assumes absolute perfection throughout the system, and the ability to produce every piece predetermined under a computerized log sawing technique at the headrig, edger, resaw and trimmer. Given the inherent imperfections in log geometry this is impossible under operating conditions. Management is, however, provided with an absolute quantitative value against which actual operating performance can be compared. Serious differences between actual and theoretical recovery should prompt an immediate investigation of wood fiber losses at individual machine centers, and improve management's decision-making with regard to operational changes or major mechanical revisions around the new technological advances in high-precision machinery, and electronic scanning devices and tallying systems.

Part 2 by Gene Neely

The trimmer performs three basic functions in a sawmill:

1. It squares the ends of the lumber.
2. It reduces the lumber to a standard length.
3. It removes defects to raise grade.

The trimmer, like other machine centers in a sawmill, is a point of fiber loss due to mismanufacturing. In fact, losses at the trimmer can occur in both the sawmill and the planer mill. It would be appropriate at this point to quote detailed studies supporting this statement. These studies, however, are not readily available. I can quote figures like $350; this is the per-shift loss that one company found in each of two mills in which it measured the loss due to trim error. Another company states that trim error represented 7 percent of the volume of lumber trimmed to remove defect, with an estimated 30 percent of the production being trimmed for defect removal. This means a little over 2 percent was lost on total production. These figures are not surprising; assuming a 16-foot piece of lumber, a 2-foot error in trim means a 12 percent loss in recoverable product for the piece. Not only does trim error mean less recovery, it means a loss of premiums on longer length lumber.

Figures for one mill are not, however, generally applicable to other mills. The errors occurring at the trimmer depend on the size and grade of logs placed into the mill. They depend on the type of headrig used, the ability of the edger operator to perform his function properly, and of course they depend on the ability and resources of individual trimmer operators.

Causal factors

Controllable—and thus significant—losses experienced at the trimmer are the result of improperly using the trimmer to remove defect or upgrade lumber. It is an attempt to remove defect in the form of wane, knots, rot and perhaps shake that results in controllable losses. The trim errors result from a number of interacting factors:

1. There is the combined difficulty to visually quantify defects.
2. The complexity of the decision.

3. The speed at which trimming decisions must be made.
4. Perhaps, to some degree, operator apathy.

For a sustained period of time, the trimmer operator cannot accurately and rapidly quantify the wane present on each piece of lumber, and then actuate a saw that trims the product to proper allowances based on lumber grade. Not only is measuring of defect difficult, the operator must rapidly and repeatedly analyze the lumber and make decisions in the one- or two-second decision window that exists at the trimmer table. The human tendency is to be safe, rather than sorry, and this leads the operator to remove a little extra wood—2 feet extra—producing an acceptable product. The questionable section goes into a conveyor for rapid conversion to chips. Once the trim end is converted to chips the decision is unquestionable.

Loss reduction

There are a number of steps that can be, and have been, taken to reduce losses at the trimmer. Operator training and monitoring is a common approach. The company I mentioned earlier with a $350 per shift loss established a training program and was able to reduce losses to approximately $55 a shift. There was, however, a problem with training; it didn't stick. After a period of time the losses again climbed to $150 to $200 a shift. Operator training must continually be refreshed, and the operator must be monitored. Continual training is possible, but monitoring of results is difficult. You could, as one individual by chance did, run the trim end conveyor past your office window, but actual measurement of trim loss due to mismanufacturing is difficult, even on a controlled study basis.

A second possibility is to have the operator trim long when in doubt, delaying the decision to the planer mill. Of course this costs in kiln capacity and may create problems in the planer mill itself. If you are selling green, the "trim long" approach will probably lower your over-all grade recovery. Another potential solution is the use of automatic controls that can rapidly and repeatedly make the trim decision based on programmed decision rules. This implies scanning the lumber ahead of the trimmer.

Trimmer control

Given existing sawmill design, lumber must be scanned as it is presented transversely at the trimmer after the lug loader. The scanner and data processor must then be able to quantify an entire piece of lumber in about one-tenth of a second, assuming a piece 4 inches wide. Scanners and processors that can handle all types of defects at this rate are not yet developed. If the automation is limited to strictly quantifying wane, however, the lumber can be scanned as it is presented transversely. By quantifying wane, I mean the amount of wane in relation of lumber thickness, width and length can be measured for analysis by programmed logic. The programmed logic can analyze the wane based on grading rules that permit, for example, 25 percent of wane in thickness, 25 percent of wane in width and 25 percent wane in length.

North American Controls has developed a scanner that can measure wane, and one channel of this scanner has been tested. To make an applications prototype would require mounting 50 to 100 of these scanners on a bar and placing them over the trimmer table. Figure 15.2 shows the trimmer control system. There is a profile scanner to measure wane; an encoder and lumber-present sensor to track lumber through the scanner and trimmer; and the operator console permits entry of grade and species. The computer processes information from the various sensors and operators entries to control the position of the ending fence and the saws. A sprayer permits optional grade marking of lumber as it leaves the trimmer. As the computer trims each piece of lumber, a lumber tally is maintained in memory. Should the operator wish to override, the computer will tally according to the operator's selection.

We are presently conducting market research to obtain more data on trimmer losses. This data will determine what portion of trim losses can be recovered with automatic control and whether recoverable losses justify the cost of the system. In the meantime we have laid out a trimmer tally system, which is really the information portion of a trimmer control system.

Lumber tally system

Log scalers have become relatively commonplace and are becoming an accepted measure of mill input. The purpose of a lum-

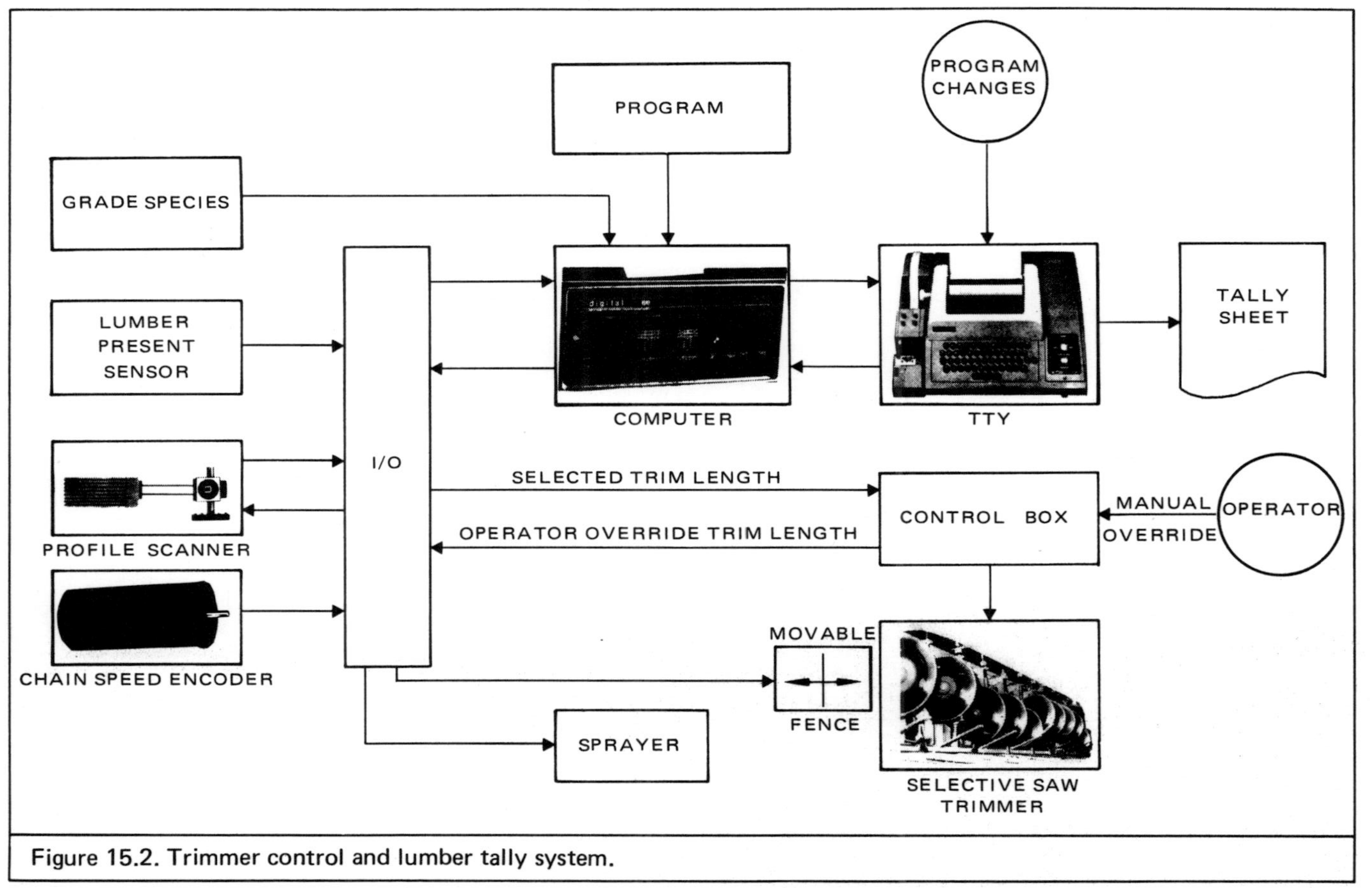

Figure 15.2. Trimmer control and lumber tally system.

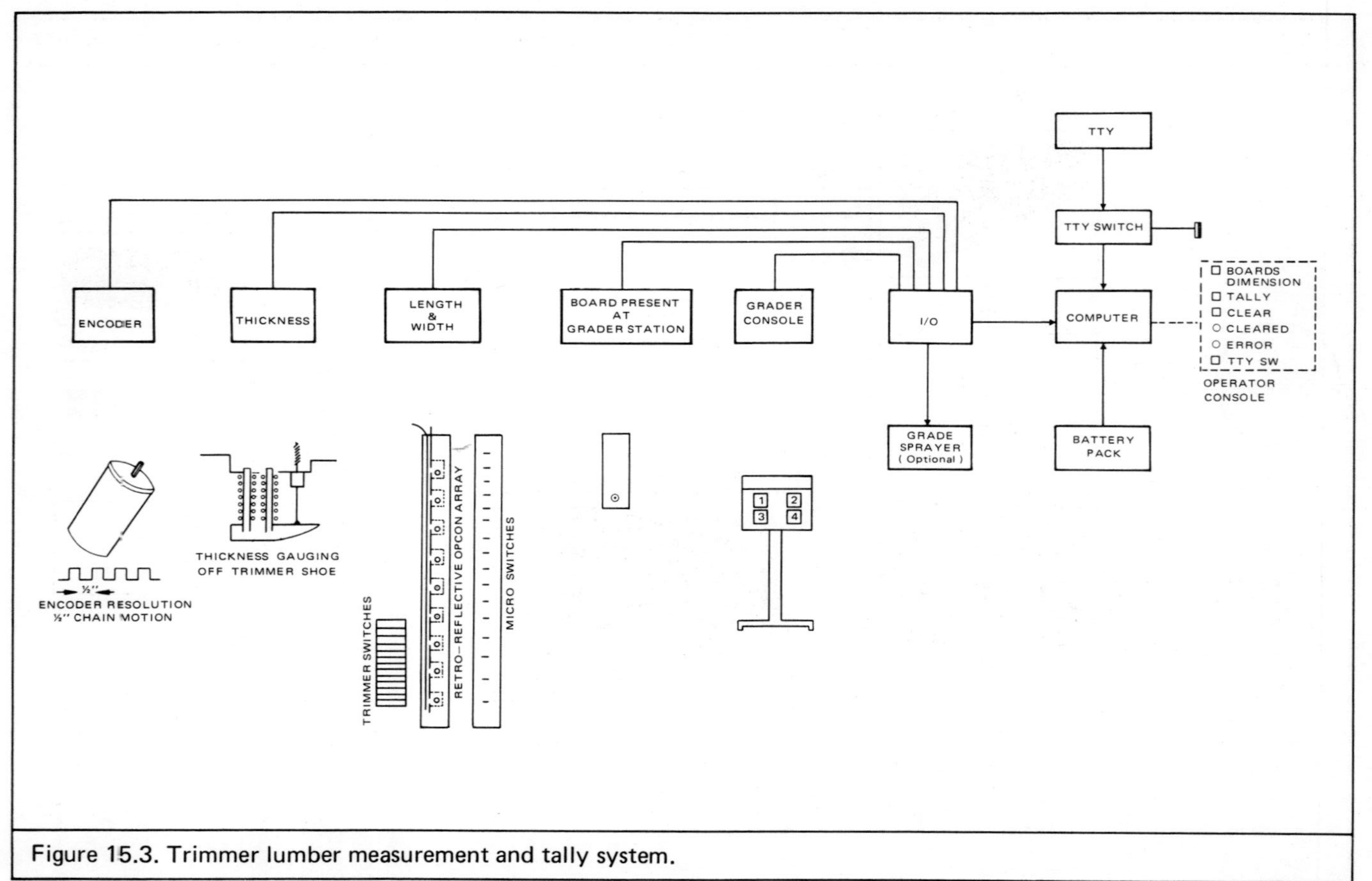

Figure 15.3. Trimmer lumber measurement and tally system.

ber tally system is to measure the dimension of each piece of lumber produced and provide a daily or shift production report by lumber thickness, width, length, grade and species if required. There are two forms of the lumber tally system: one is a hard-wired version that provides only a piece count, the other is a minicomputer-based system providing tally data in board foot measure and offering a number of other forms of information. Figure 15.3 outlines one configuration of a trimmer tally system. Again, lumber position is monitored with a pulse generator on a lumber-present sensor. A shoe is illustrated as a lumber thickness gauge, but there are noncontact sensors similar to the profile scanner described above for measuring lumber thickness. Lumber length and width can be determined from optical or micro limit switches. If the tally system is to be installed immediately following the trimmer, the computer can determine lumber length by monitoring trim saw selection. The operator's console permits entry of grade, which can be sprayed or stamped under computer control. The teletype (TTY) permits listing of the lumber tally.

The information provided by the lumber tally system has three primary uses:

1. Inventory control
2. Production control
3. Sales management

For inventory control the production reports will provide the necessary data to maintain a perpetual inventory of lumber in the yard or in-process to the planer mill by lumber dimension and grade. The production data may be punched to paper tape and entered into a time sharing system to prepare detailed reports, or a certain amount of the lumber inventory information may be maintained in the memory of the computer version of the lumber tally system. The tally report can be used by the mill manager as an accurate record of shift production, without the confusion of accounting for partial loads left on the green chain at the end of a shift. As the industry moves toward computer control of primary breakdown machines using predetermined sawing solutions, it will be possible to relate actual recovery to the theoretical recovery that was programmed at the headrig. The system will also have the capability to combine log scale input data with the lumber tally so the recovery factor for the mill can rapidly be determined on a shift or daily basis. Log costs can also be allocated to the individual products produced based on the mill's equation for allocating joint costs, if it has one, and this can provide the basis for determining contribution margin by product dimension. Contribution margin is the difference between sales price and the variable cost of production; that is the number you want to optimize. As a sales management tool the lumber tally report can be used to verify or isolate any shortages or overages in a "production-to-order" operation. If lumber is being produced for inventory the sales department will be provided a rapid breakdown of the product mix available for sale.

In summary, trimmer control packages can be developed that make trimming decisions based on quantitative measurements of wane. These systems will provide a means of reducing losses due to trim errors. The lumber tally system that can be incorporated into the trimmer control system, or that can stand alone, will provide necessary information for inventory control, production control and sales management.

16

Continuously rising temperature drying

Dr. Dallas S. Dedrick, Lumber Drying Consultant
Irvington-Moore, Portland, Oregon

One of the most memorable experiences of my younger days came on the eve of receiving a graduate degree. Our research director, Dr. James Newton Pearce, gave a farewell party. He asked us what we wanted to do in research when we got out into the world and became active researchers on our own. We told him a number of exotic things we thought would be interesting, things we thought needed to be done. Largely, they represented spin-offs of the research we had already defended for our degrees. Dr. Pearce said, "No, you shouldn't do that. You should study boiling and freezing and diffusion and temperature—things like that." We replied, "Dr. Pearce, you're off your rocker. These are the tools that we already know all about. These are the tools that we use." He contended, "No, these are the things you should study."

In the decades that have come and gone, people did study boiling; as a result, it is now possible with fractionation to get pure liquid never dreamed of before. The study of freezing has resulted in zone melting processes that give us the pure crystals that are the basis of the transistor industry and many, many other things. People studied diffusion; out of these studies came chromatography, membrane technology and completely new vistas in medicine.

If Dr. Pearce were alive today, I am sure that he would say, "By all means, go study boiling and freezing and things of that sort." If this same man had been knowledgeable about or interested in the lumber processing industry, I am sure he would have added to the list.

Dr. Pearce would have included such things as moisture content and its measurement; he would have asked just what happens when you create a new surface from a piece of wood by sawing, cutting or abrading. Just what really does happen? He would have been interested in shrinking and swelling and I am sure he would have been interested in the process of drying.

Lumbermen practicing an art?

Things such as moisture content, shrinking and swelling represent the common tools of the lumber manufacturing industry. They are generally considered to be understood and practiced in their most effective manner. Yet if you were to ask 10 experienced saw filers how best to sharpen and temper a saw, you would probably get 10 widely differing answers (unless some refused to talk because they considered their methods secret and proprietary). And if you asked 10 experienced kiln operators for the best schedule for drying West Coast hemlock, you would very probably receive 10 different recommendations. Of course, the saws sharpened by all of the filers perform in the mill to produce lumber, and each drying schedule results in a seasoned product acceptable to plant management.

However, the existence of such a wide variety of procedures for accomplishing a single purpose strongly suggests that manufacturing steps are being practiced as an art rather than a science. A better understanding of the basic principles involved, and the application of these principles in practice, should result in more efficient and economical manufacturing.

Several years ago, Weyerhaeuser Co. decided that, as a large-volume producer of wood products, it was imperative that the company learn and practice the most basic principles of the various manufacturing steps. Responsibility for making fundamental studies was given to the research division. It was my extreme good fortune to have been the leader of the research division team that was charged with finding out more about the fundamental principles of lumber drying. Our team had no preconceptions; we were free to test any hypothesis that appeared to have merit. Our management was cooperative and extremely patient. We learned much. We caught glimpses of productive areas for future investigation that time did not permit us to study immediately. We answered some questions. In the process of answering them, we formulated many, many more questions—questions that we were unable to even conceive of when the study started.

Since my retirement from Weyerhaeuser, I am no longer as actively engaged in purely investigative research. I am happy to say that the work there continues under the very able direction of Dr. Gilbert Comstock and Kendall Bassett and that I remain on their staff in a minor consulting role. I spend the major portion of my time as a lumber drying consultant to Irvington-Moore. But, if time could be rolled back 20 years and the opportunity presented, I could think of nothing more challenging, more opportune—nothing I would rather do—than start a basic study in lumber drying.

As Dr. Pearce would say, "Study and restudy those fields that you think you understand. Study them objectively and carefully, because the potential values are limitless."

Experimental kiln

A major portion of our research study dealt with the measurement of air and wood temperatures and with the calculation of mass and energy balances and transfer rates under widely differing conditions in an experimental dry kiln. This experimental kiln had a holding capacity of approximately 1,000 board feet of lumber in a package 8 feet long and 8 feet wide.

Thus, end effects were minimized and drying characteristics across a conventional commercial charge were duplicated. Air flow and dry-bulb and wet-bulb temperatures were precisely controllable and load weight was recorded automatically every 15 minutes. Each board was weighed and graded, both green and after drying.

Some of these studies, which extended well over a decade, were incorporated into what is now known as the constantly rising temperature (CRT) process.

The concept has demonstrated sufficient uniqueness and utility to have merited the assignment to Weyerhaeuser of basic United

States and foreign patents for the process. Irvington-Moore has purchased world-wide exclusive license to introduce and install the process.

Just what is the CRT process? What are the processing and economic advantages of using it? Does CRT offer advantages over conventional processes?

Simply described, CRT is a technique of controlled heat transfer that results in an approximately constant rate of moisture removal. This mechanism permits use of substantially shorter kiln time than conventional processes. If adequate intake and exhaust venting are available, CRT is a low-temperature process—started at temperatures only slightly above the outside ambient temperature—with fast drying rate. The strain on boilers or other heat sources during the conventional initial heat-up is eliminated. More importantly, drying during the initial stages (when the wood is most subject to physical and chemical degrade) takes place at less damaging low temperatures.

Air needs reduced

For many applications, the quantity of circulating air may be substantially less than is conventionally provided and a maximum of one air direction reversal is recommended. By far the most important economic bonus contributed by CRT is its inherent capability to yield higher product values than other processes that use the same kiln residence times. Valid data prove that the increased product value amounts to several dollars per thousand board feet, which completely overshadows other factors such as capital and operating costs.

The economics of lumber drying reduces to one simple principle—production of maximum product value with minimum dollar expenditure. Product value relates to the quantity and quality of the finished lumber. Maximum product value is attained when the minimum amount of footage fails to reach the shipping shed because of out-of-specification moisture content, warpage, splitting, planer and other losses and when grade distribution of the dried product is highest.

Processing cost includes items such as energy expenditure, handling, maintenance and capital cost per unit volume of lumber

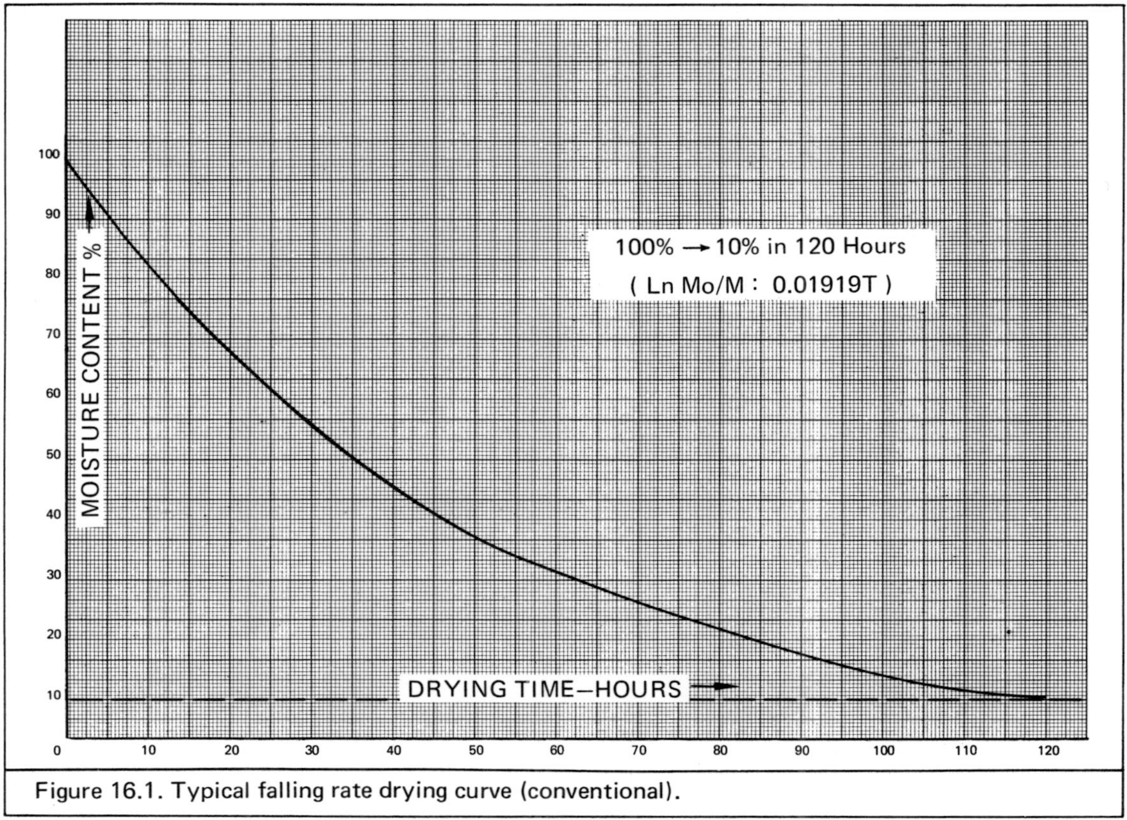

Figure 16.1. Typical falling rate drying curve (conventional).

dried. The first reaction of most people is probably that the most important element of drying cost is kiln residence time. Faster drying schedules permit more lumber to be dried by a single kiln, thus reducing capital cost per thousand. This is, indeed, an important factor and, frankly, was foremost in the minds of the Weyerhaeuser drying team when the study was begun.

Drying rates charted

Conventional drying is characterized by a falling rate of moisture loss. Moisture content is plotted on the Y axis in Figure 16.1; the X axis shows drying time in hours. The curve has been calculated to show a charge of lumber being dried from 100 percent mean moisture content to a mean of 10 percent over 120 hours. This very closely approximates what would happen under a conventional constant-temperature process. Here is what happens: Early in the process, the drying rate is quite high. In this particular case, it amounts to about 1.9 percent for the first hour. Drying rate decreases to about 0.15 percent in the 119th hour. Thus many hours—even days—may be needed to remove the last few percent.

Figure 16.2 demonstrates why the rate falls. All of the energy that is transferred from the circulating air to the wood is not used to evaporate water. Rather, a very sizable portion of that transferred energy is used to heat the wood and its contained moisture. If there is a constant dry-bulb temperature (illustrated by the top horizontal line in Figure 16.2) and a wet-bulb temperature maintained as shown (lower horizontal line), the wood does not remain at the wet-bulb temperature. Rather, it very quickly rises above the wet-bulb temperature and asymptomatically approaches the dry-bulb temperature.

The quantity or rate of transfer of energy from air to wood is determined by the temperature difference between the air and the wood. That temperature difference is the "driving force" for heat energy transfer. If the wood continuously rises in temperature, the dry-bulb temperature remains constant. The difference between the dry-bulb temperature and the wood temperature constantly decreases. The driving force is, therefore, decreasing. The quantity or the rate of energy transfer is decreasing and finally becomes

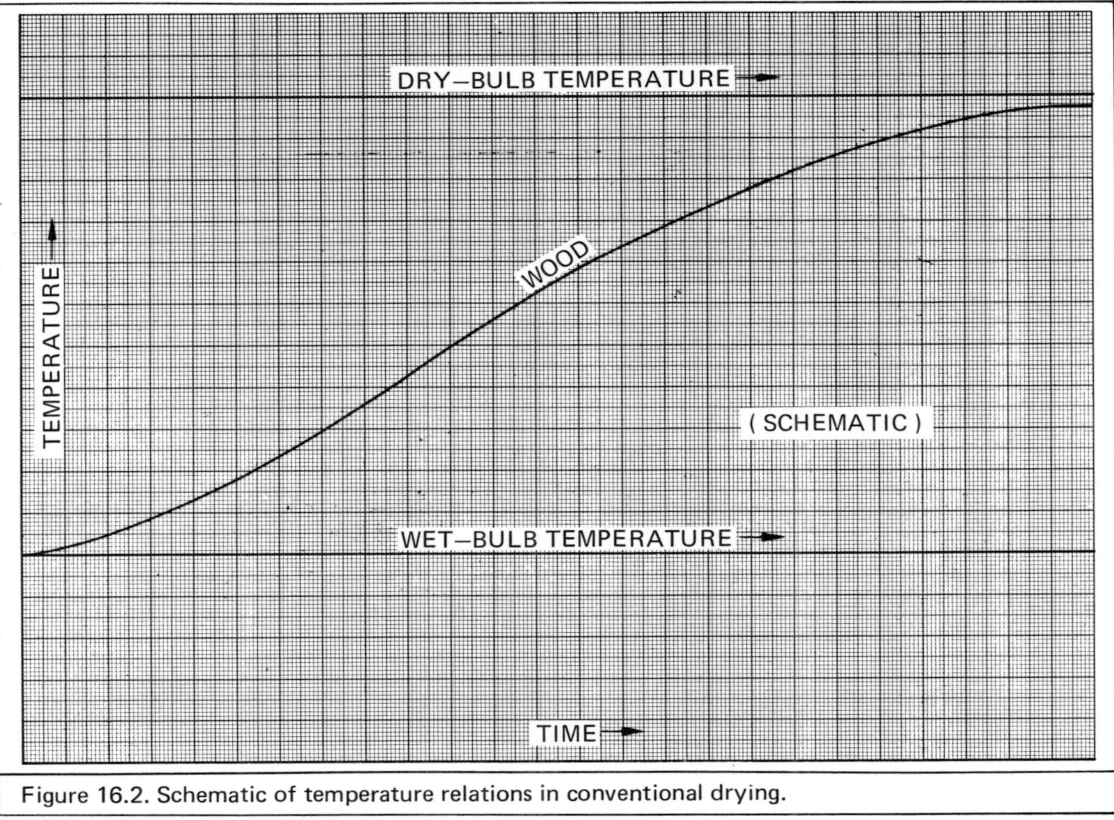

Figure 16.2. Schematic of temperature relations in conventional drying.

negligible indeed. So this, we believe, is the primary reason for the falling rate of drying in conventional processes.

Constant differential

What can be done about the falling rate of drying? Figure 16.3 shows what we attempted as a hypothesis. We reasoned that if the wood temperature (indicated by the middle line) is going to increase, then we must keep the dry-bulb temperature above the wood temperature, thereby maintaining a constant gradient between air temperature and wood temperature.

The wet-bulb temperature, of course, must remain below the wood temperature to insure that drying occurs all the way across the board, or the load. So we reasoned that if we maintained a constant air-to-wood gradient, we would achieve a constant rate of energy transfer from air to wood and, therefore, approximately constant-rate drying. This actually takes place in practice. Although absolutely constant-rate drying is not achieved, it is very closely approximated.

Drying rates contrasted

Figure 16.4 demonstrates the difference between falling-rate and approximately constant-rate drying. The solid line reproduces the curve from Figure 16.1—the 100 to 10 percent fall in drying rate in 120 hours. Figure 16.4 calculates value for constant-rate drying. It assumes that the same rate achieved for the first hour in the falling-rate drying was maintained until 10 percent moisture content was achieved. Thus, by maintaining the drying rate established during the first hour—by holding it constant—the 10 percent level is reached in 47 hours instead of 120 hours. If only two-thirds of the initial rate were established and maintained, time required would be in the neighborhood of 70 hours. If only half of the initial rate were established and maintained, drying would be completed in 90 hours. At a factor of only four-tenths of the initial rate, we still come out with slightly shorter kiln residence than is required for the calculated falling-rate method.

By maintaining the initial rate shown in Figure 16.1, we would

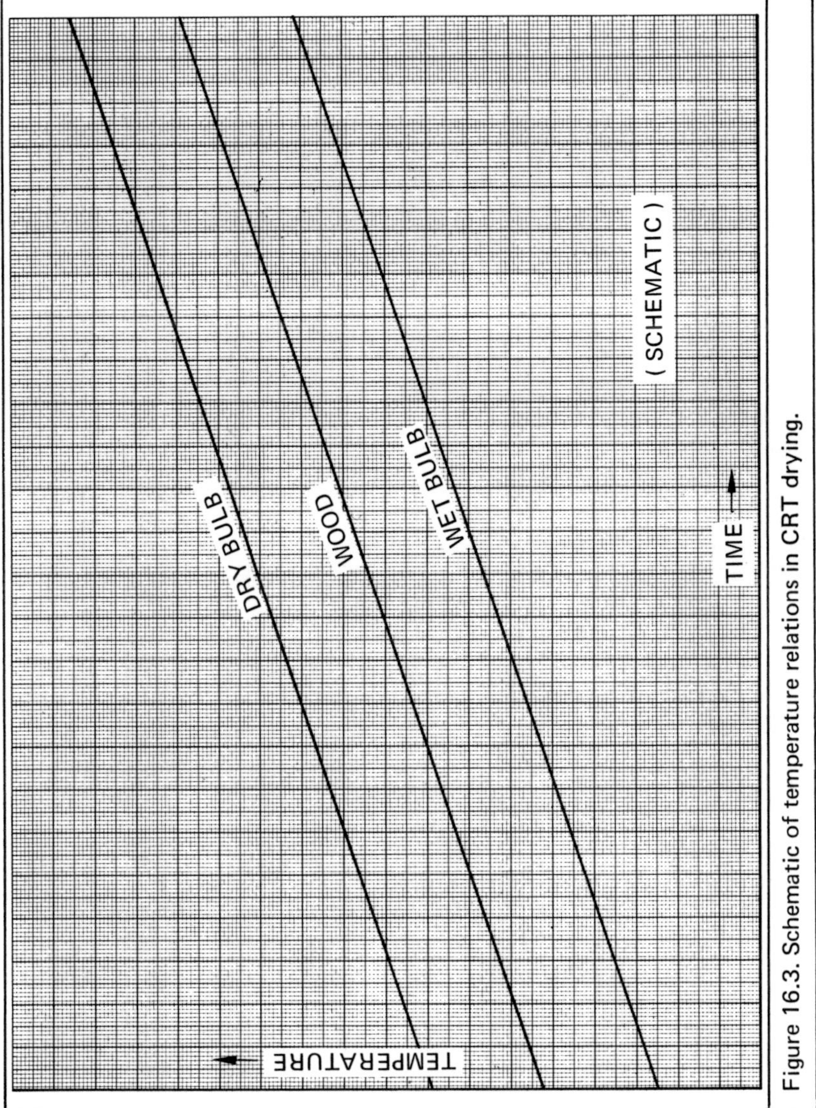

Figure 16.3. Schematic of temperature relations in CRT drying.

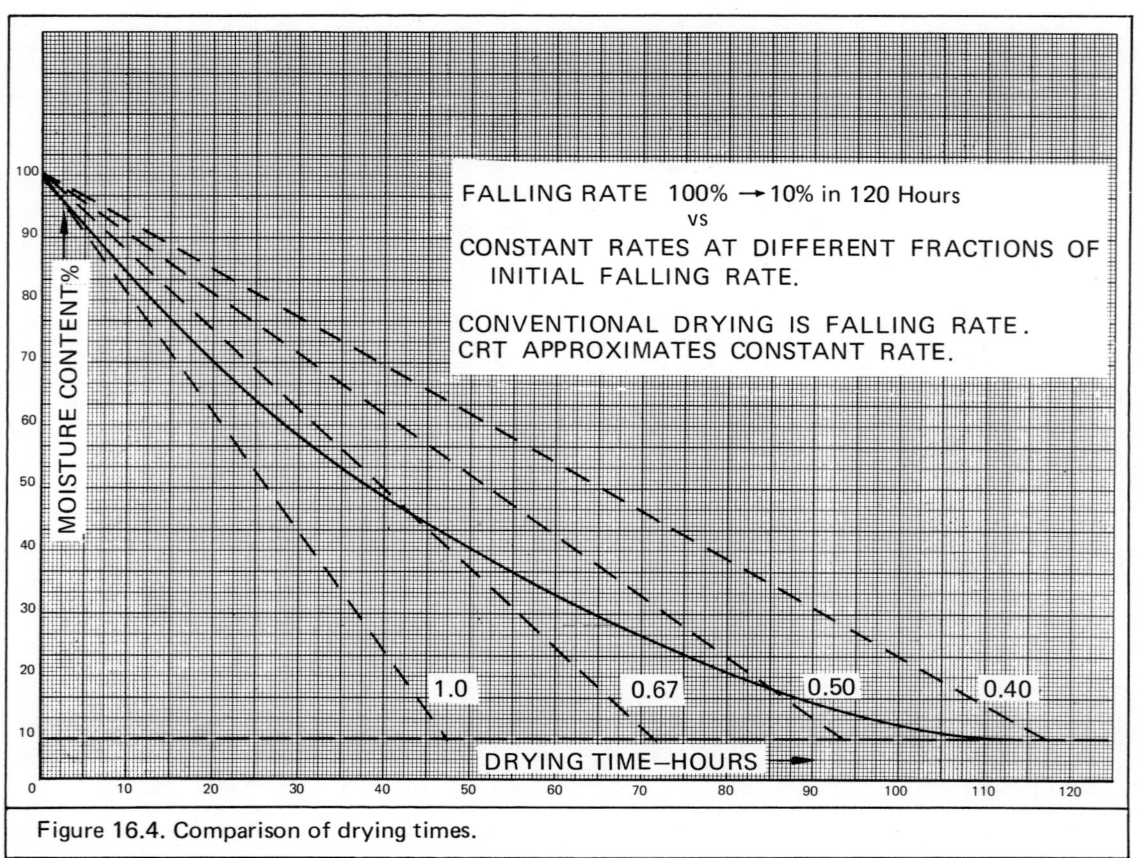

Figure 16.4. Comparison of drying times.

end up at 47 hours. What sort of a falling-rate drying process would be required to achieve that same time? Instead of the 1.9 percent that we talked about originally, it comes down to the neighborhood of 5 percent—a very, very, very fast drying rate.

Kiln time cut

It is inherent in the substantially constant-rate drying (CRT) process that low rates of moisture loss result in shorter kiln residence times than conventional falling-rate processes that employ higher initial rates of moisture removal. This is important, not only because of reduced kiln residence time, but because these low drying rates are used during the critical early stages of drying. Improved lumber quality is achieved primarily by a combination of shorter drying times and slow initial drying rate.

Obviously, if dry-bulb temperature is to be increased continuously over a substantial time, the initial temperature must be quite low to avoid reaching excessively high temperatures at the end of the drying period. Consequently, CRT drying starts at low temperature. Depending upon ability to achieve and maintain the required low-humidity kiln conditions necessary for fast drying, typical CRT starting dry-bulb temperatures are in the 90- to 120-degree range. Several examples of dry-bulb temperatures in the 60- to 80-degree range are on file.

One of the thrills of my life is to have seen recently—it was cold outside—plumes of steam rising from a kiln drying ponderosa pine where the dry-bulb temperature of the internal entering air was 106 degrees. The plumes of steam continued to rise, of course, throughout the drying period. Incidentally, no spray steam was used; we do not recommend the use of spray steam at any time during the process, other than for conditioning, if conditioning is desired. Thus, initial removal of moisture not only takes place at a lower rate but at a lower temperature than in conventional drying.

Degrade dropped

The characteristics of the CRT method, individually and in combination, contribute to an economic bonus substantially larger than that represented by reduction of kiln residence time and the corresponding increase in kiln throughput and reduction in capital expenditure. This bonus is the significant increase in product value that has been demonstrated for all items for which statistically valid data are available. Carefully controlled tests involving replicated, large-scale comparisons, using matched samples, have yielded unexpectedly high dollar advantages for CRT drying.

When compared with a 230-degree dry-bulb/180-degree wet-bulb schedule, sometimes erroneously called "high-temperature drying," CRT drying yielded increased product values of more than $6 per thousand board feet for both Douglas fir and southern pine dimension stock when dried to the same moisture content in substantially the same time.

When compared with conventional drying, CRT also showed improvement in product value despite the fact that the kiln residence time for conventional drying was two to three times as long as for CRT. CRT-dried lumber is characterized particularly by lower incidence of both warping and checking than lumber dried at the higher temperatures or by conventional methods.

Increased product value is due to a combination of (1) reduced planer loss and (2) maintenance of a higher dry-surfaced grade distribution. Qualitatively, CRT-dried lumber is also characterized by lighter color and more satisfactory surfacing than comparable items that are dried by either conventional or the higher temperature processes.

Strength loss less

Higher temperature drying has been proven significantly deleterious to the strength of some lumber items. There seems to be no question that this is true of Douglas fir. Statistically valid data on strengths of Douglas fir dimension stock show that CRT-dried lumber has strength properties equivalent to conventionally dried material and substantially higher than matched samples dried in the same time to the same final moisture content by higher temperature processes. Data on strengths are less positive for other items. There is argument in the literature as to whether the higher temperature processes are harmful to all species. Evidence is not positively convincing.

In summary, CRT drying is a method of controlled heat transfer that results in drying times competitive with higher temperature schedules, but that yields product values equivalent to or higher than those achieved by conventional processes. It approximates constant-rate rather than falling-rate drying. It is highly flexible in application. Substantial benefit may be attained with minimum modification of existing kilns. Additional performance and economic advantages accrue as more nearly optimum kiln design is approached.

As a crude rule-of-thumb, modified existing kilns may be expected to CRT-dry typical lumber items in one-half to three-fourths the time required for conventional methods. New kilns, designed with CRT performance in mind, will perform substantially better—we have no idea how much better.

The CRT process has never been tried on large-scale lumber tests with kilns capable of pushing the process to its limits. A kiln is being designed to see just how far we can take the process. We are thinking of providing heat, air and venting capabilities to dry items from 100 percent green moisture content to 10 to 12 percent moisture range in 12 to 16 hours. Will it work? We do not know; we have never pushed CRT drying to these limits. The only way we can find out is to submit CRT drying to tests of this nature.

Energy transfer process

We frequently misuse the term *schedule*. CRT is not a schedule; it is a method of controlled heat transfer. Conventional drying, accelerated drying and high-temperature drying are all methods of controlled heat transfer; they are not schedules. They are simply rates of application or transfer of energy processes. There is no such thing as a CRT kiln; there is no such thing as a high-temperature kiln or an accelerated kiln. These are simply kilns that have the capability of transferring energy at varying rates.

The following are characteristics of the constantly rising temperature (CRT) method of drying lumber.

1. CRT is not a schedule; it does not require a CRT kiln. It is a method of controlled heat transfer, as are other kiln drying processes.

2. CRT is characterized by a constant rate of moisture loss, rather than a falling rate.
3. CRT provides gentle, low-temperature conditions during the early stages of drying when degrade potential is the highest.
4. CRT application requires kiln residence times that are competitive with higher temperature processes, but yields product values equivalent to, or higher than, those achieved by conventional processes.
5. When CRT is used commercially, drying times vary with equipment, species and other conditions. Douglas fir dimension is being dried many places in 30 to 40 hours. I can verify 24-hour schedules in some places. Douglas fir 2-inch clears are being dried in 50 to 60 hours. Engelmann spruce is dried routinely in at least two mills in British Columbia in 24 to 30 hours. Hemlock dimension is being dried in 30 to 45 hours. Southern pine dimension is being dried in 28 to 32 hours. Limited experience with maple and birch (furniture-grade material) indicates two to three days less than conventional time.

CRT is probably not the ultimate in lumber drying. I would like to think that it is not. Continued research into drying's mechanism and kinetics will undoubtedly result in improvements in both technology and economics. However, the concept does represent distinct advantages over past processes. Now, since the CRT process is proprietary, I am sure that I will be excused for not describing its application in sufficient detail for all mill operators to rush to their plants and start practicing the art.

DISCUSSION PERIOD

DICK JUST, Boise Cascade Corp., La Grande, Oregon: Is the CRT process affected in the early stages of drying or is the transfer of energy influenced if you start with frozen lumber?

DEDRICK: The temperature of the wood is quite likely to be more nearly constant as long as you have an ice phase present. There would be a tendency, probably, if you were raising the air temperature, for the air-to-wood temperature difference to increase somewhat until the ice phase had disappeared.

JUST: Would you recommend a warm-up period before starting to raise your dry-bulb?

DEDRICK: In limited experience, we have not used preliminary warm-up. Certainly preliminary warm-ups have not been used in British Columbia with Engelmann spruce. Whether preliminary warm-up would have merit, frankly, I do not know.

DON BAACK, Crown Zellerbach Corp., Portland, Oregon: Are energy requirements the same for CRT and conventional drying?

DEDRICK: As to the actual energy involved in the evaporation of the water concerned, you have a specific amount of water to be vaporized. Heat evaporization is substantially constant over a very wide range of wood or water temperatures. Admittedly, it is slightly higher at the lower temperatures than it would be nearer, say, the normal boiling point of water. There is, then, starting at very low temperature, a minuscule increase in energy requirements for evaporation. However, add the fact that your kiln is operating at low temperatures so radiation losses are lower because of the smaller gradient between the kiln and the outside atmosphere; there is less energy involved in heating the ambient air taken in. I think that it more than outbalances the small difference in heat evaporization.

TOM RADTKE, Commander Industries Inc., Red Bluff, California: Does it make any difference whether you are using a high-pressure or low-pressure boiler system? Would it make any difference if you were using a double-track dry kiln instead of a single-track dry kiln?

DEDRICK: The answer to the first question is absolutely no, unless your heat supply just happens to be inadequate to maintain the rate of temperature rise or the rate of vaporization that you calculate you should have. Basically a BTU is a BTU. The idea is to get it out of its source and into the air for transfer. No, so far as steam is concerned, it makes no difference at all.

If the double-track kiln is such that the air is moving in series through both loads, you will, of course, be picking up double the amount of moisture, and dilution air requirements will be greater. So long as you re-establish between the loads enough temperature to have drying all the way across the load, use of a double-track kiln makes no difference. Double the moisture pick-up requires twice the dilution air to re-establish the entering air conditions for a given quantity of air being circulated.

BEN JONES, Longleaf Industries Inc., Columbus, Georgia: Are these kilns available today?

DEDRICK: Yes. You can buy them today.

At the takedown chain, dry lumber is spot checked for moisture content.
Photo courtesy FOREST INDUSTRIES magazine

17

Surfacing dry lumber by sanding

Sherman Kirchmeier, Plant Manager, Louisiana-Pacific
Corp., Oroville, California

Abrasive planing is being discussed more and more in our industry. Because of the outside pressures being applied to the forest industries, wood utilization is becoming synonymous with survival rather than a subject for further consideration.

Precision sawing is already a reality to most of us. Any system that will complement this program or aid in achieving additional profits certainly must be examined by our industry. Abrasive planing could be one such system. Let us examine its advantages and current status in our industry. There are two types of abrasive planing systems available: (1) the batch system and (2) the in-line system.

Batch system

We have installed the batch system in Louisiana-Pacific's plant at Oroville, California. The system involves a dual line that includes a pair of two-headed edge planes, two accumulating lines and a four-head Timesavers abrasive planer or sander 52 inches wide. It is similar to the type used in particleboard and plywood plants.

The batch system operates in this way: Boards are first passed, one at a time, through the edge planer. Next they fall onto accumulators that align the pieces for entry into the sanding machine. This machine will take up to 12 2x4s at once. If the stock removal is realistic, the feed rate possibility of the wide belt abrasive planer will be at least 125 feet per minute. This calculates to 1,625 board feet per minute, a speed certainly advantageous in our industry.

In-line system

The in-line system is the type Louisiana-Pacific has installed at Samoa, California. As its name denotes, the in-line system entails abrasive planing of the top, bottom and both sides—one board at a time. It offers flexibility—profiling knives and saws can be mixed with the abrasive heads. It would be a natural system for conversion of existing planing mills. The abrasive planer can be installed in the area where the knife planer presently sits, using no more room. It might utilize the existing offbearing system. Where space is tight, or where established plants plan to renovate existing surfacing departments, this system would be a likely choice.

As part of the system at Oroville, Louisiana-Pacific developed a sectional roll feed specifically for dimension lumber. Rolls are segmented and have steel on the outside with rubber in the middle to allow for any unevenness in the 2x4s coming in.

52-inch sander

Advantages of the 52-inch sander used in this system are close tolerances, no knot damage, no grain tear-out, improved finish, high production, maximum yield and low maintenance.

The in-line system at Louisiana-Pacific's Samoa plant has a typical moulder infeed with four heads and a profiler in back.

Now let us list the advantages and disadvantages of abrasive planing itself and examine the merits of each system. If there is a consistent heavy cut necessary with high speeds, abrasive planing is not economic. Abrasive planing requires good tolerances. However, if a plant is sincere in precision cutting and/or avoiding downgrade due to the chipping-out or tearing of grain, this is an ideal system. At Oroville, we can shoot for much closer cuts with no fear of losing grade. This has been especially lucrative in our vertical grain shop program. Shop lumber that contains large knots is never chipped or torn. This enables us to pick up even more grade—at least one grade and often two.

Allowance of 1/64th

The less stock you give the abrasive planer, the more it excels. Feed speeds increase, abrasive belt life increases and grade recovery remains the same. Many times we have taken dried lumber that had only 1/64 inch to clean up—both sides. We have been able to burnish that board enough to maintain the grade. I am sure you cannot do that with a conventional knife planer. Of course, I must admit that we are not quite able to saw to those tolerances.

When we started the Oroville plant, our thickness variation was substantial. We had only one accumulator loading our Timesavers abrasive planer. It could handle the batch boards at only 80 feet per minute. Our finished grade and our belt life did live up to expectations, but the footage output did not. Slow feed speeds required us to run the finish department six days, three shifts.

However, we have learned to control thickness by better sawing and drying techniques. On our kiln-dried lumber, we are setting our ganged edgers at 1 21/32 inches, running at 300 feet per minute, and are consistently attaining high accuracy throughout the length of the boards. Our consistency will run between 98 and 100 percent, plus or minus 1/32 inch. We have also learned enough about our dry kilns to achieve maximum efficiency. We now can easily control the moisture content without getting more than the allowable skip.

Planer output double

As a result, abrasive planer production has doubled. In fact, we have since added another accumulator, another set of side heads, and we are running the finish line five days, two shifts. To be more specific, the side head planers have the advantage of maintaining speeds of 750 feet per minute or faster, while the abrasive planer handles this production with a feed of 150 feet per minute or better. This plant has run as much as 321,000 feet in nine hours. We run 250,000 board feet consistently every eight hours, using

14 men per shift. After we make some changes, we hope to reduce the number of men by about four.

Belt costs have been reasonable. Costs do vary somewhat, but usually stay in the neighborhood of 50 to 75 cents per thousand board feet. This is low, everything considered. Belts are easy to change; they are relatively inexpensive to maintain. Downtime is minor compared with a planer.

Crooked lumber handled

Crooked lumber can be handled easily. We can hold tolerances in thousandths of an inch from side to side and from end to end without any skips. This is possible because of the unique feeding systems, pressure plates and fixed heads of this abrasive planer.

Grade is recovered virtually 100 percent of the time when you include shop and finish material. When you consider the high production and large wood recovery on construction grades, belt costs are very economical.

We are using a conventional open coat type of abrasive in 24 and 40 grit. The 3M Co. assures us that soon it will supply some new, more durable materials. These materials should cut costs and make the system more versatile.

We hope that the new abrasives will allow us to run our small amount of green lumber completely wet. Thus far, we have had to dry the perimeter of green lumber before passing it through the abrasive planer. This has not really been as great a disadvantage as we first thought it would be, but we are still looking for an abrasive planer that will run completely wet lumber. The 3M Co. hopes to have such an abrasive product for us soon.

Hopes fulfilled

We are pleased with the batch system at Oroville. It has more than lived up to our expectations and it has allowed us to use our sawing techniques fully. Anyone who is serious about cutting lumber accurately should consider abrasive planing.

Let us review the new system at Samoa, California. This is the first in-line abrasive planer in our industry. Its main purpose was to size and finish finger-jointed stock.

This particular abrasive planer, manufactured by the Kimwood Corp., includes a moulder-type feedwork and two top and two bottom 15-inch-wide abrasive heads. It is followed by a knife moulder that is connected to the sander. This moulder, which can utilize our existing profiling heads, is necessary in manufacture of varied products, including fascia boards, S4S, and so forth.

Glued stock upgraded

The primary purpose of the machine is to eliminate chipping-out of the finger-jointed stock and tearing of grain around architectural boards. Prior to installation of this system, the Samoa mill was downgrading about 40 percent of its glue joint stock to B grade by chipping out the offset glue joints with the knife planers. The abrasive planer, as planned, has completely eliminated this; now we have 100 percent recovery. The only B grade we experience now is due to an improper glue joint.

The unique sanded surface, though it appears different from that of a knife surface, has been accepted by the market. Our sales people have told us that painters prefer the sanded surface over the conventional knife surface. There is no question that paints and sealers adhere to this surface far better than to the tightly surfaced knife surface. Painters have appreciated this benefit.

Disadvantages listed

Moving from the benefits of the system at Samoa, I would like to mention its disadvantages. This system has had many imperfections, and still has. None of us involved in the installation actually realized what some of those imperfections would be. Kimwood Corp. has had to make major on-the-spot modifications, and is making more. The 3M Co. learned quickly that its conventional abrasive materials were far from adequate. It had to expedite a backing that up to this time no marketplace had seen. When the installation was planned, everyone involved thought that four abrasive heads would be enough to give us the necessary production and finish—a conclusion drawn from experience in abrasive planing. The conclusion proved incorrect.

Size, finish separately

It was quickly learned that when you run at high speeds and require a medium to fine surface on redwood, you cannot both dimension and finish with the same abrasive heads. We learned that as soon as you apply much pressure to fine grits, they load up and burn prematurely. We thought we had four heads to dimension when it turned out we had only two. This is not enough to remove 0.125 inch of stock maintaining feed speeds above 300 feet per minute.

This problem resulted in lower production and higher belt costs than anticipated. Progress has been slow, but we are gaining. The 3M Co., the Kimwood Corp. and Louisiana-Pacific feel that we are nearing complete success. When this is achieved, there is a strong possibility that we may add two more heads. This will add more versatility to the system and provide better economics. We recommend that anyone considering this type of system buy not less than six or eight abrasive heads.

Improvements coming

Abrasive planing is certainly in its infancy; there are problems to be resolved. However, there is so much activity in the field that we know today's methods will be improved upon tomorrow. This was true of the system at Oroville three years ago and will certainly be true of the system at Samoa. New equipment and products will be introduced. As improvements are made we, as an industry, will be the beneficiary. Louisiana-Pacific obviously feels that abrasive planing will save wood. I recommend that everyone, no matter what type of sawmill he has, explore abrasive planing.

Photographs of Oroville operation

The following photographs show the system in operation at Oroville, California. All lumber at Oroville is pulled to length. It has come out of our double-end trim. We have two side heads. Lumber comes out of the trimmer with one line above and one below—one going to each side head.

In Louisiana-Pacific's Oroville system, all lumber is pulled to length. Rough lumber moves from double-end trim to side head.

A workman at Oroville is breaking the lumber down. It has been fed through the side head machines.

Lumber comes out of the side head machine and drops onto an accumulator.

One of the two accumulators where stock is batched at Oroville. The 2x4s go through the sander 13 at a time.

Lumber comes out of the sander and drops onto the grading table.

Lumber passes in front of the graders. If material arrives as it should, three men can grade 100,000 board feet each without any problem.

Lumber volume coming down to the pullers. Approximately 70 percent of the volume goes over the end and is not pulled.

DISCUSSION PERIOD

TOM FLEMING, American Forest Products Corp., San Francisco, California: What make is your lumber tally? You mentioned 320,000 feet—is that on an eight-quarter count or a six-quarter count? What is the approximate cost of your back system?

KIRCHMEIER: Because there have been so many changes, I do not think I could really give a true cost figure. It is on an eight-quarter count.

BILL GREENE, Potlatch Forests, Inc., Lewiston, Idaho: First, does lumber have to be trimmed before it goes to the sander? Second, have you run four-quarter stock? Third, what modifications were needed to handle sander dust as opposed to shavings and how costly were they?

KIRCHMEIER: First, lumber does not have to be trimmed before it can go through a sander; this is no problem at all in feeding.

Second, we have not run any four-quarter in the past two years. Third, regarding handling the sander dust "without blowing up your plant"—well, without blowing it up, it is a problem. We have had several fires. The only cost it has been to us has been in loss of production time. The sander dust is a problem—increasingly so in white fir, which is about 60 percent of our cut. We have not completely solved the problem we have with sander dust. Hopefully, we might have a system to take care of it in the future, but for now there is still a problem.

PETER HEISER, Heiser Industrial Tooling Co., Seattle, Washington: You said you were cutting about 95 to 98 percent within 1/32 inch. What type of machine and blades are you using to achieve that?

KIRCHMEIER: The machine is a gang edger. We have a top arbor machine built by Ukiah Machine & Welding. The guides we are using are of the Thrasher type.

HEISER: But you also said, in talking about your second mill, that you were having to allow 1/8 inch to plane but you seemingly were able to cut your lumber to 1/32 inch. Why do you have that much to take off when you apparently can saw so precisely?

KIRCHMEIER: This concerns our Samoa operation and I would have to put the question to the people at Samoa. I can only vouch for what we are doing at Oroville. The 0.125-inch stock removal is at Samoa.

UNIDENTIFIED: Did you mention the horsepower required?

KIRCHMEIER: No, but each head at Oroville has 150 horses on it. The machine that is at Samoa has 100 horses.

UNIDENTIFIED: Is it correct that the sander takes considerably more horsepower?

KIRCHMEIER: Yes.

18

Setting the mill to minimize maintenance

Bob LaBelle, Sawmill Engineer, U.S. Plywood Division, Champion International, Bonner, Montana

No one can tell you a quick, easy way to set up a good plant maintenance program. This can be done only by someone familiar with the people as well as the physical plant.

A sawmill, being an in-line type of operation, is prone to breakdowns that greatly curtail production or completely shut it down. Some mills have less lost time than others, but all mills—at least all mills I have been in—have some lost time due to mechanical failure. The purpose of a good maintenance program is to keep frequency and seriousness of lost time as low as possible.

Always set your goals to the Utopian ideal that your maintenance program will eliminate lost time. But always bear in mind that if you save 10 minutes a day, per shift, you are gaining five days of production per year.

Maintenance continuous

When setting up a maintenance program, keep in mind that maintenance is a continuing thing. We have all heard people say, "I wish I had new equipment to work with so I would not have to fight this broken-down junk." But new mills that are not constantly and well maintained soon become that worn-out piece of junk we are fighting now. Sawmill maintenance must be taken seriously or the results can be costly.

The methods for setting up a maintenance program vary with the individual mill and people. But certain fundamentals remain more or less constant.

So let us begin with the selling of a proposed maintenance program. The first person you must sell is yourself. If you are not completely sold on planned maintenance, forget about it, because you must have complete confidence in the program yourself before you can sell it to others in your organization.

When selling a program to management, remember that very few programs can show immediate or spectacular results. They must depend on results accumulating over a period of time. The quickest way to have management lose faith in your program is to promise unrealistic results.

Keep hopes realistic

Let the boss know in dollars and cents what he can probably expect in increased cost until the program is fully established and operating. Then let him know what he can expect in overall long-term savings. Stress and prove such things as reduction in lost time, reduction of maintenance crews, reduced material cost and reduction of maintenance overtime. Remind him that reduction of lost time means increased production and possible reduction in production overtime hours.

After management is convinced that a maintenance program is essential, the supervisors must be sold. They must be convinced that the program will increase production or that it will make their jobs easier. Give them an opportunity to help with your program. They can probably offer constructive criticism as well as help you with ideas or indicate areas or functions that you have overlooked. It certainly will not hurt the supervisor's ego or decrease his co-operation if he knows he helped to get the maintenance program off the ground.

Millwright support needed

Millwrights will go along with the program—once they are convinced that a good maintenance program can make their work easier and that the superintendents are behind the program. The millwright carries out the actual functions of the program, so do a good job of selling here. One man not sold is like a bad apple in the barrel.

Production people should not be left out. Any program you develop should be explained to them. They should know the reasons for the program and how it will affect them and their jobs. If it is at all possible, try to involve production people in the program, even if it means a little extra work for them. The extra work could mean, possibly, a weekly operator's machine report.

Responsibility, though apparently shunned by some workers, is something most people take pride in, providing it is dealt out in small doses. Making a man responsible for the maintenance of a machine or system and letting him know that maintenance decisions on this equipment are his, will usually result in immediate enthusiasm. Results are usually overwhelming. It must be explained that a priority system and checks and balances will be used to determine what work is to be done when, where and in what order.

Rotate responsibility

People, especially those with any ambition, usually dislike doing the same job every day or working in an area they know they will be stuck in for a long time. This natural restlessness lends itself to periodic rotating of responsibility. Rotating maintenance people not only gives you a built-in informal training program but also can be rewarding in that almost everyone—there are exceptions—believes he can do a better job than his predecessors. He also would sooner be dead than let someone show him up after he has completed a job. This gives him added incentive to do his best.

One word of caution: some people must be convinced that taking their responsibility from one area and moving it to another is not affecting their job security, but, in fact, is increasing it by giving them experience in the overall operation.

Keep records lean

Records, next to people, are probably the most important item when setting up and operating a successful maintenance program.

Since we all want any program to be as simple as possible, yet to be effective, records must be set up to require as little daily effort as possible. Requiring too much bookwork of maintenance people has probably shot down more maintenance programs than any other thing. It is wise to stick strictly to basic, necessary records. If you want to be more elaborate, of course, that is your prerogative in planning the maintenance program.

Records basic to a successful maintenance program include the following:

1. Files of manufacturers' information.
2. Electric motor and gear box records.
3. Lost-time records.
4. Preventive maintenance records.
5. Maintenance checklists.
6. Lubrication records.

Let me give you some idea of how to compile this information and what to include in each category.

A manufacturers' file, consisting of machine description, drawings, parts lists, and so forth, is absolutely essential. This file should be kept where maintenance people have free access to it but where they do not have to run half a mile to look at it. Provision should also be made for a desk or table to lay out drawings and prints so no part of the file is removed from the area and not returned. If a manufacturers' file exists on the property, it should be reviewed and updated. I am sure that if information on a machine is nonexistent or incomplete, the manufacturer will send you information on request.

Index cards help

Electric motors and gear boxes recorded by machine or area on index cards will save much guesswork and lost time. Electrical motor information should be a duplicate of the data printed on the motor plate. Figure 18.1 gives an example of the type of information we record.

Gear boxes should also be recorded, using the information on the nameplate. Sprockets, chains, belts, couplings, clutches or oth-

MOTOR RECORD

DEPARTMENT Sawmill	MACHINE NO.	MOTOR NO. 358
MFG. Westinghouse	SERIAL NO. 2-4V1416	OPERATES Conveyor by Gang Motor

TYPE Code G	H.P. 10	RPM 1735	VOLTS 220/440
PHASE 3	AMPS 26/13	FRAME 256-UY	CLASS
MODEL NO. ACFC	CYCLES 60		TEMP. RISE 55°c
PUR. ORDER NO. S#4V1416			STYLE

BEARINGS. FRONT END 1603468

BEARINGS. REAR END 1449627

SHAFT DIA.	SHAFT LENGTH	COUPLING
LUBRICATION EVERY DAYS	GENERAL INSPECTION EVERY DAYS	MONTHS

GEAR HEAD MOTORS OR GEAR REDUCERS

MOTOR INTEGRAL WITH GEAR	MOTOR COUPLED TO GEAR
GEAR MFG. Western Gear	SERIAL NO. 210
REDUCTION 56.12 TYPE	MODEL 256PG4316
SPEED OUTPUT SHAFT 30	SIZE OUTPUT SHAFT
TYPE COUPLING OUTPUT SHAFT	GEAR SIZE

CONTROL EQUIPMENT

CONTROL NO.	MFG. G.E.	NEMA SIZE 1	NEMA ENCL.
TYPE CR106CO	DUTY FWD	FUSES	VOLTS 600
PHASES 3	AMPS 13.5		BREAKER 40A. Cat.#I36040

358   Sawmill – Conveyor by Gang Motor

Figure 18.1. Data on electric motors and gear boxes may be recorded by machine or area on index cards.

er equipment used in conjunction with a motor or motor gear box should be carefully recorded. If the nameplate is off the motor box or gear box, try to get all information possible, such as rpm, shaft size, and so forth. Do this ahead of time and systematically file the information. It can save hours in an emergency.

Lost-time records

Lost-time records are probably one of the best tools we have. They enable us to determine trouble spots. Properly used, they provide a basis for preventive maintenance as well as an indicator of mechanical, electrical, material flow or personnel problems. Figures 18.2 and 18.3 show examples of these. It is a shame that so often lost-time records are kept only as a buck-passing tool. In recording lost time, minimize detail and avoid long descriptions. If trouble is pinpointed and briefly but accurately described, details can be gathered when needed.

These records should be kept in one central spot in order of occurrence. A large looseleaf binder makes a very good record book. It is easy to thumb through, allows sheets to be added and is compact. Again, keep lost time by machine or area so that trouble spots can be easily seen. Make this a perpetual record and it can be used to determine such things as belt, chain, motor or gear box life.

Plant supervisors should compile these records, since they are in the best position to deal with daily problems. But all personnel connected with maintenance should have access to these records. It must be emphasized that lost-time records, if kept and not used, are worth less than the paper they are written on.

Build on experience

Preventive maintenance records should be built around experience—your own, that of the equipment manufacturer and information gathered from lost-time records.

When you determine a preventive maintenance item, it should be recorded and scheduled so as to allow it to be checked off when done, then transferred to the next scheduled date. A diary-type book, such as a DayTimer, is useful for this purpose; records

#1 Set works   LOST TIME

Date	Cause	Day O	Day M	Day E	Night O	Night M	Night E
1/10/72	Broken set chain		15				
1/15/72	Broken wire on hammer dog limit						5
1/16/72	" " " " " "			6			
1/17/72	Reset 4/4 cam		4				
1/30/72	Broken air line on pressure switch					10	
2/10/72	Broken wire on carriage junction box						10
2/11/72	Broken air line on pressure switch		11				
2/13/72	Broken air line on pressure switch		8				
2/20/72	Adjust dial					5	
3/10/72	Broken air line on pressure switch (night shift replaced pipe with tubing)		15				
4/1/72	Broken air line (log hit it)					13	
4/15/72	Adjust dial chain tightener		3				

Figure 18.2. Lost-time record on the setworks.

Edger   LOST TIME

Date	Cause	Day O	Day M	Day E	Night O	Night M	Night E
1/10/72	Hot Saw — Sliver in guide	5					
1/12/72	" " " "				5		
1/15/72	Feed roll chain broken		10				
1/15/72	" " " "					10	
1/16/72	" " " "		7				
1/16/72	Sprocket off feed roll					8	
1/16/72	Feed roll chain off (night shift chain replaced & lined sprockets)					6	
1/17/72	Edger plugged — crooked board				7		
1/20/72	Edger plugged — cant too thick	12					
1/30/72	Broken saw tooth (change saw)				10		
3/3/72	Edger plugged split board	5					
3/15/72	Feed roll chain broken — stick caught in it					10	
3/20/72	Edger plugged — cant too big	10					
4/15/72	Edger plugged — crooked board	5					

Figure 18.3. Lost-time record on the edger.

of this type can be most rewarding over the long haul.

Immediate results are not spectacular, but when you start comparing your record with last year's lost time on a specific item and find you have cut lost time in half by replacing a belt periodically or tightening down a cover on schedule, you will feel pretty good.

Maintenance scheduling

Scheduling of day-to-day maintenance becomes routine when a maintenance checklist or inspection sheet is developed for each man or shift. Unexpected repairs will definitely have to be made between shifts and some day-to-day maintenance and inspection will also be needed then.

When scheduling work during operating hours, include overhauling of cylinders, valves, and other parts. This time is also good for fabricating new equipment, rebuilding equipment that has been replaced but could be used as a spare and, last but not least, seeing that all equipment needed for downtime jobs is collected before it has to be used. One word of caution—when scheduling such maintenance work during production hours, be sure the job is in the production area and that it can be dropped if it is necessary to answer an unexpected maintenance call.

Lubrication records essential

Lubrication records, schedules and checklists are essential if you are going to keep lost time down and a maintenance program running smoothly. Lubrication schedules have a twofold purpose. They assure that each lubrication point is serviced on schedule and they make the oiler, if properly trained, your best maintenance inspector. The lubrication checklist decreases the possibility of the oiler missing a lubrication point. It also allows the supervisor to spot-check the oiler as far as schedule and quality of lubrication are concerned. Your petroleum products supplier will usually set up a lubrication list showing proper lubricant, schedules, and other information. Using this list, it becomes relatively simple to make out your own checklist.

Any system you develop must have at least two qualities if it is to be successful in a sawmill: (1) it must be versatile in that it can change with people and equipment; (2) it must be simple to follow and simple in its bookkeeping. A successful system cannot become so completely routine that it does not require day-by-day follow-up as well as continual analysis and updating of its individual components.

People the top ingredient

People are the most important ingredient in a successful maintenance program. Keeping them informed of scheduled work is as essential as seeing that they have materials and tools with which to work. With fairly competent maintenance personnel, proper maintenance tools, a few simple records and a desire to accomplish a smoothly running maintenance program, lost time can be reduced and profits increased. Remember, keep it simple, keep people involved, follow through and remain versatile.

DISCUSSION PERIOD

CLIFF JACKSON, American Forest Products Corp., Martell, California: Bob, do you feel that an oiler should be a millwright, or do you feel you can pick just about anyone to do this job?

LA BELLE: If I had my wish, he would be the best millwright I had on the property.

19

Can culls be cashed?

H. B. "Dutch" Gram, President, & John Gram, Vice
President, Gram Development Co., Portland, Oregon

Part 1 by John Gram

There is really no single or simple or correct answer to the
question, "Can culls be cashed?" In this presentation, we can barely scratch the surface of the problem. I could be flip and perhaps
even accurate in saying, "In today's market, anything can get
cashed in." But many sawmillers know, or I hope will better understand through this presentation, that there are many wheels
within wheels in trying to cash culls.

The chapter on Gram Development Co. will cover, in general,
two main, interrelated areas: (1) some information and observations on the breakdown sawing of culls for lumber and chips; and
(2) some thoughts on the complex concept of complete forest
utilization.

First, let's clarify some definitions. *Substandard* logs are all logs
not graded as commercial or merchantable sawlogs because of size
or defect. Substandard includes utility grade, which is less than
one-third merchantable but at least 50 percent chippable in
content. Substandard includes the so-called cull or wood log,
which is less than 50 percent chippable, so it is not even utility
grade. Substandard also includes all logging residue slash process-
able, possibly, for some product. How much substandard material

exists? (We will not count the sawlogs you all get that you claim are not even a good utility grade.)

On each average clearcut, 120 tons of residue are left. Two-thirds of this residue is over 3 inches in diameter. A recent residue study indicated that more than 20 percent of an average clearcut stand remains as potentially utilizable material. Statistics compiled from sales on a Gifford Pinchot ranger district from 1968 to 1973 show that 24 percent of the volume taken out was substandard—either utility or a cull grade.

The U.S. Forest Service says that the residues left in the Douglas fir region amount to 50 percent of the total annual wood consumption of the pulp, paper, lumber and plywood mills in Oregon and Washington. And that 50 percent does not take into account the substandard material now being taken out as utility or a cull grade.

In the past 10 years, there has been much study and talk about reduction of logging residues. But there has been very little real action.

The Forest Service Experiment Station and such private firms as Gram Development Co. have researched all around the problems. There are mills that saw utility logs either as a full-time diet or when the market or mood strikes them. Others have developed peeler blocks from utility—some with success, some without. But the bulk of the logging residue is still unused. At the Sawmill Clinic, we have talked about the timber shortage; we have talked about increased yield and increased recovery. People look at logging residue as a symbol of waste and a misuse of natural resources. They feel that the loggers and sawmill people are not taking proper care of the environment. But there are several other factors beside the raw material shortage and environmental factors that make us look more closely at substandard utilization. Reasonably clear land is necessary for prompt reforestation, so that the slogan "Our Only Renewable Resource" is an actuality and not just talk. Technological breakthroughs are coming in many different areas—in equipment, automation, optical scanners and computers. Research continues in extruding, segment gluing, pressed boards, and in bark-chip separation methods that would permit chipping without barking. Then there is the Forest Service sale contract factor, which could be the biggest subject of all—the topic for an all-day seminar, as a matter of fact.

Gram Development Co., at an old circle headrig mill using a Rosser head barker and a bull edger that has since been dismantled, tried to cut a blend of low-grade saw logs and utility grade, alternating between fir and hemlock, and produce enough lumber and chips from these low-priced logs to turn a profit. The 1969 to 1971 market was not the ideal time to run an old bucket of bolts. We learned the shoulds and should nots of low-grade log handling. More important, we learned that the problem is much bigger and broader than just handling the logs. It is more complex than just trying to balance reduced production and reduced yield with lower raw material cost.

A little more than a year ago, Gram Development did an interesting operations study for the Forest Service Experiment Station. It combined much information that we had learned ourselves or gathered from other outside sources active in sawing utility logs. We marked arriving utility or cull grade logs and saved the Bureau scale slips with the log line number marked. On the log deck, before bucking, Forest Service people examined each log, estimated the chip and lumber content percentage, premeasured the small and large diameter and the length and recorded all this information with the log number. After barking, each of the 8- to 12-foot log sections was end painted in rotating colors, was remeasured and re-estimated for lumber and chip recovery. Each log segment was timed on the headrig, with notations for downtime and the reason for being down, such as "the log fell apart." As the lumber produced from each log segment was end color coded in color sequence, it was possible to list by size, length and grade each board produced from each bucked log segment. The test lasted two long days and included logs with all degrees and types of defect. Diameters were from 4 to 52 inches. At the end, we knew the exact total gross and net utility, Bureau Log Scale, the yard-delivered cost, the exact lumber production by size, item and grade, all the dollars pertinent at that time, plus the wet chip weight delivered to Boise Cascade and the number of bone dry units and dollars that we got out of it. We also kept track of the sawdust and fuel and, of course, the lumber production comparison to each log scale, net and gross.

In today's market, being able to sell the sawdust (which we

were not able to do at that point) and not worrying too much about the extra rot and foreign material in the chips, we could have made money sawing utility. But on the long pull, with the equipment we had, it was a "no go" situation.

As a rule of thumb, if you take the substandard or utility log with an approximate 20 percent dockage from gross to net, the economic breakeven point is getting a minimum of 50 percent of the net log scale in lumber—or, if you prefer, 40 percent of the gross, with at least 2½ units of chips per lumber thousand, plus a little gravy from the sawdust, hog fuel or barkdust sale. The key to getting lumber out of a substandard log is to get the length of lumber where it is in the log and quickly hog or chip the rest. To do this you need equipment to handle the substandard log. The entire system layout must take into account an entirely different pattern of occurrence frequency in the defect location. The equipment that you use must be consistent with the objective. And you need a superintendent, foreman, sawyer and crew who do not always think overrun but think about getting what is there and then getting rid of the rest quickly. Crew and supervisory psychology should not be dismissed lightly. The concept of getting what is there is alien to many sawmill people. Sawmill thinking is generally production per hour, per day. Budgets are all tied in at volumes in overrun.

To handle substandard logs you need approaches, not only in system and equipment but in people training, that get the most out of the low end material. Logs are the only material that we have with which to work. With the good logs, there are also poor logs.

A summary of what we have learned in several years of experimentation is that there is good lumber quality in some quantity when you saw substandard logs. You must have specialized equipment and system layout to handle the logs; you must have the crew working with you to find the lumber in the defect; and you're really dead in a poor lumber or chip market. Market stability is of paramount importance.

This is what we feel just might be a workable approach to the problem. Rather than bring in the substandard log—in actual practice, it is usually the better substandard utility log—and put it through a mill operation geared to cutting the merch log, a better concept would be to establish a residue operation to process *all* substandard material, from the better substandard utility through the poor cull and into the very marginal logging residue that still has some usable product. At very least, the substandard utility log should be sawn by a specialist mill.

Ideally, a complete timber utilization program might be set up as follows: Substandard logs and usable logging residue would be transferred to a central processing location. There, processing might include:

1. Sorting by species, size and condition.
2. Whole log barking and hogging of the bark residue.
3. Bucking and sorting by peeler or chipper.
4. Shipping the bucked and barked logs, or peeler log segments to the manufacturing site.

Or further processing could occur at that point:

1. Chip, and ship the chips.
2. Hog the remaining material.
3. Use the balance of the hog for steam electric generation.

A steady, long-range and large volume of substandard material obviously must be available over a long period for any such central processing unit. Processing and transportation would probably need subsidization as the project would be very marginal except under the most stable and best market conditions. The subsidy might have to be a cooperative program between industry and local and federal government units, in some type of nonprofit corporation. The subsidy could take a variety of forms. The program is actually a log merchandising system including varying manufacturing processes designed and operated with the objective of getting the most value out of substandard material and converting everything left—after the last marketable product has been extracted—to some type of energy.

The independent, nonintegrated sawmill, buying public timber sales, must soon face up to a new priority factor—his sawlog supply is directly related to substandard utilization. This relationship involves both supply availability and the price of the sawlog he wants to cut.

Part 2 by H. B. "Dutch" Gram

From the foregoing discussion, it is quite evident that the subject is knotty, with many facets. In 1964, Gram Development Co. completed a study for the government entitled "The Economic Feasibility of the Utilization of Logging Residue." I regret that in the ensuing nine years very little progress has been made. But here are a few general observations that I should like to make.

The first is in regard to present logging practices. Let us put aside the everyday use of equipment and consider that contracts

Large solid material was left on the ground after slashburning a Douglas fir clearcut in the Lewis River district of the Gifford Pinchot National Forest. Above, clearcut after the merchantable logs had been removed.

are generally predicated on a per-thousand basis. Therefore, the normal logger is interested in volume of logs, primarily commercial logs. He is not very much interested in anything below the commercial log in grade. In other words, he is interested in moving a large volume of merch logs. Over the years, we have tried a number of things that we thought might help take care of waste in the woods. We thought at one time that pricing per acre would be an answer. It has not proved so. We have tried yarding unmerchantable material (YUM logging). We discovered that the cost allow-

This photograph, taken in the same cutting block of the Gifford Pinchot National Forest as the picture on facing page, shows residue after slash had been broadcast burned under U.S. Forest Service supervision.

ance that the government put on YUM logging was not sufficient. It has gone up continually since, and no one knows where it will end. But when this material is brought to the landing, the logger does not know what to do with it. So, it seems that this material— good, bad and indifferent—must be taken away to some central location. Oregon State Senator Bob Smith has indicated that it is necessary for us to recognize that we must grow timber. It has always been disappointing to me that the industry has not made a bigger effort to inform the general public about forest operations. A standing joke tells of the congressman from New Jersey who

Large volume of unmerchantable logs with some lumber, some chips and much hog fuel potential remaining in the Gifford Pinchot National Forest.

was astonished to find out that you had to cut down a tree to build a house. So, we are in a peculiar situation out here in the West where we know about our own problems but the public does not. And unfortunately, every nickel the Forest Service gets goes into the general pot. And every nickel it spends has to be budgeted out. If we ran a business like that, we would probably be pretty well broke.

In analyzing the salable products from substandard material, we are actually interested in what chip demand is and what it is likely to be. Of course, there have been studies made on that. It would

The cutting block shown on facing page after it had been broadcast burned under supervision of the U.S. Forest Service.

appear from the talks at the Clinic that we are getting more lumber out of the log. But I would imagine we also are getting fewer chips. If that is true, I would think that there ought to be some additional demand for chips from other sources such as substandard material. It has been estimated by some of the authorities that by 1977 Japan will have doubled its requirement for chips. Our own manufacturers in the United States are well projected. Their projections indicate that there is no difficulty in selling chips. We cannot expect them to bear heavy additional prices—I would not suggest by any means that the price of chips is going to go up. Naturally, there is some reaction when chips get short, but I would not expect that to be any answer to the problem of cleaning up the woods.

The cost of transportation enters into the picture very definitely. This is another reason why it is desirable to establish some type of operation near the logging.

Assume that we decide to use all the substandard material and that lumber and chips have been recovered. What remains is the really bad stuff that has to go into fuel—hog fuel, preferably. For the Douglas fir region, the Forest Service estimated it would take 50 processing stations to clean up the residue.

The Forest Service estimates that about 12 million dry tons of logging residue are produced annually in the Douglas fir region. This has nothing to do with mill residue. Mill residues are pretty well taken care of by the plant owner. The problem is what to do with this raw material that has no market.

We have been hearing much about energy shortages. However, I do not believe that this should be thought of in terms of a national emergency on energy. It can be thought of in terms of local energy shortages. Think back to the 1920s and 1930s when nearly any mill of any size produced not only its own electricity but supplied some for neighboring small towns. Possibly we have come full circle to the point at which we will be utilizing this residue material. But that does not even scratch the surface of available material, nor does it scratch the surface of the necessity to clear these lands.

We must get all the old big stuff out of the road if, later on, we are going to thin, fertilize and use our ingenuity to practice good forestry. The smallest central generating plant that conceivably could be competitive would be about 25,000 kilowatts. That is not of a size to be competitive with larger operations.

When we analyze utilizing hog fuel instead of coal in, for instance, the Centralia, Washington, area, we discover that we are dealing with a bulk product. The cost of transporting hog fuel is entirely out of order so far as being competitive with strip coal. It has been estimated that you would have to deliver logging residue to Centralia at 60 cents a dry ton. That is quite impossible.

What about type of equipment? Well, we are just going to have to rely on Yankee ingenuity. Maybe you are going to use a hayrack to take away substandard material at one landing. I don't know. But I believe that someone other than conventional loggers should become involved. Properly prepared, special personnel should be involved in the entire operation. The psychology here must be entirely different from getting out production. It was quite a while before I realized that there was a psychological factor involved.

Index